KT-547-982

VERBUM DEI LUCERNA

WALLACE ARTHUR

THEORIES OF LIFE

DARWIN, MENDEL AND BEYOND

PENGUIN BOOKS

Penguin Books Ltd, Harmondsworth, Middlesex, England
Viking Penguin Inc., 40 West 23rd Street, New York, New York 10010, U.S.A.
Penguin Books Australia Ltd, Ringwood, Victoria, Australia
Penguin Books Canada Limited, 2801 John Street, Markham, Ontario, Canada L3R 1B4
Penguin Books (N.Z.) Ltd, 182–190 Wairau Road, Auckland 10, New Zealand

First published 1987

Made and printed in Great Britain by
Richard Clay Ltd, Bungay, Suffolk
Typeset in 10 on 12pt Photina

For Helen,
who hates theories

CONTENTS

PREFACE

This book is aimed at anyone with an interest in the nature of life on earth, in the scientific theories that have been advanced to explain it and in the conflicts that have arisen between the scientists putting forward these theories and others – such as the 'creationists' – for whom any scientific theory of life is anathema. The book is written in as jargon-free a way as is possible, and has a 'detachable basic biology course' (Chapter 3) which can be skipped by biology undergraduates and the like, but which will serve to explain to readers without any biological background the meaning of the small number of scientific terms that are unavoidable in a book of this kind.

Modern biology, or 'life science', really started in 1859, the year Charles Darwin published his famous *Origin of Species*. In the current biological literature there are still many references to the work of Darwin and some of his contemporaries, but very few to pre-Darwinian biologists. In many ways, Darwin's 'evolution by natural selection' was the first successful scientific theory of life. When the idea of natural selection was first put forward its general philosophical implications were evident to all, not least the Church, from which much of the initial opposition stemmed. Present-day biology, however, is seen in a very different light by society, and comes across in the media of the 1980s as an increasingly practically orientated or 'applied' science spearheaded by biotechnology. The emphasis seems to be on how to do things with biological techniques, rather than on understanding the fundamental processes of life.

With the emphasis of modern biology shifted towards industrial matters, accepted biological theories, particularly evolution, have increasingly come under attack from non-scientists. Also, with the fragmentation of the present-day biological sciences into many specialist disciplines, the importance of more recent theories of life often does not filter through to the layman as readily as the importance of Darwin's theory did more than a century ago. There is thus a danger that the popular picture of life will be increasingly dominated by non-scientific views, while biologists bury their heads in the industrial sand.

I hope in this book to lessen this danger by doing three things. First, I shall explain the most important biological theories of life at a level which I hope will be readily understandable – with the aid of Chapter 3 – to a non-biologist. Secondly, I shall draw these theories together and show how they interrelate, rather than merely giving isolated accounts of the theory of evolution, the theory of heredity and so on. Finally, I shall expose the extreme creationist view for what it is – a paranoid and ill-thought-out rejection of the most central biological theory of life, a rejection which is unnecessary for a religious view of the world, unacceptable for a scientific one, and dangerous in that it sees these two views as mutually exclusive, which they are not.

One key feature of the scientific approach is that scientists admit when, in a particular area, their theories are inadequate, incomplete or unproven. This is still the case in some areas of modern biology, notably in the study of development – or embryology as it is sometimes called. I shall make it very clear, when dealing with a particular theory, whether it is an accepted one or one still subject to modification or even replacement. The fact that major uncertainties still exist in biology in no way weakens the case for a scientific approach to life, and indeed most scientists get much more excitement from these uncertainties and their possible resolution than from resting on the laurels of the well-established theories of the past.

ACKNOWLEDGEMENTS

I would like to record my gratitude to Dr Alec Panchen and my wife, Dr Helen Arthur, for reading through, and commenting upon, the typescript. I have been sufficiently pig-headed to resist making some of the alterations they suggested, but I have taken account of others, and the book has undoubtedly improved as a result.

1 INTRODUCTION

The thing about life on earth that most readily strikes the casual observer is its great variety. Humans, earthworms, pine-trees and bacteria are strikingly different and are easy to contrast in all sorts of ways. What is harder to see is that the great diversity in the structure, appearance and behaviour of different kinds of organism masks an underlying general theme of life which is an attribute of *all* life-forms whatever they look like and wherever they live. The general theme to which I refer is order and a tendency towards ever greater order. This general theme applies not only to whole organisms but also to parts of organisms and, going in the other direction, to collections of organisms in time or space. Moreover, all other general themes of life that have been proposed, such as evolution, are subsidiary to, and essentially components of, the general theme of order.

I should stress at the outset that order is a meaningful concept despite the difficulty of defining it. It is instructive to attempt a definition, and if you feel the need of one I suggest that you make the attempt, because I am not going to supply a categorical statement of what I mean by the term. I hope that, by the end of the book, the nature of order in living systems will be clear. For the moment I shall content myself with the observation, with which few readers will disagree, that a typical living room or bedroom with constituent furniture, carpets, curtains and so on is more ordered than a random collection of empty drawers, the items that were in them and everything else in the room thrown together in a haphazard way. Associated with this mental image, we can say that the chambermaid is an agent of order, the thief an agent of disorder.

Not only is the ordering tendency of life a feature that sets it aside from inanimate objects in general, but indeed there is a theory – the Second Law of Thermodynamics – that tells us that the universe as a whole is proceeding to more and more disordered states. Does this mean that living systems contravene, fail to obey or disprove in some way this basic law about the physical universe, as some authors have implied? The answer is a categorical 'no' and the reason for this is very simple. Basically the Second Law applies to what the physicist calls a 'complete energetic system' – that is, one in which there is an energy source (such as the sun), an energy 'sink' (deep space) and some 'intermediate system' through which the energy flows from source to sink. If we take any one of these components alone, there is nothing in the Second Law which tells us that it cannot proceed to greater degrees of order, so long as any such localized ordering processes, all added together, do not outweigh the trend in the opposite direction that characterizes the rest of the universe. Thus our earth may provide a local countercurrent to the general rundown going on around it, and this countercurrent is running its course in a type of system to which the Second Law simply does not apply.

In fact, the reference to the 'earth' is a little misleading. Actually, most of our planet is entirely devoid of all forms of life. This must at first seem a rather odd statement, given that everywhere we can think of – from the arctic tundra to the centre of London or New York to the prolific rain-forests of the Amazon – life-forms are easily found, albeit in varying form and abundance. But we must remember that our outlook is biased by the fact that we are surface-dwelling creatures, whereas planets are solid entities and not just soil/air interfaces. If a hypothetical 'observer' were to travel along the radius of our planet, from its core to the edge of the outermost atmosphere, he would traverse a distance of several thousand miles. Along this rather uncomfortable journey he would encounter living organisms from a few metres below the surface to a few hundred metres into the air – a tiny fraction of the total distance travelled. In fact, the earth's biota (all its plants, animals and micro-organisms) is to be found in a very thin envelope called the biosphere, a cross-section of which we have just described. As with the meteorologist's stratosphere, ionosphere, etc., the biosphere is not a sphere at all – rather it is a sort of circular envelope on the inside of which is a sphere with which we are unconcerned and outside of which lies the rest of the universe. It is the biospheric envelope that contains all of the earth's life-forms, and this region is thus the geographical domain to which the theories described in this book relate. However, it may not be the only such domain. Almost certainly none of the other eight planets of our solar system have biospheres, but there may be others further afield which do.

Let us return for a moment to the laws of physics. The fact that living systems individually, and the biosphere as a whole, do not contravene these laws does not of course mean that they are adequately explained by them. Thus, although the ordering trend of life does not violate the Second Law of Thermodynamics, that law in no way explains why the biosphere *should* be a zone of intensive evolutionary (and other) ordering – it only *allows* it to be so. For an explanation of the actual ordering mechanisms we must look elsewhere. This is why there is clearly a need for a systematic study of life – that is, a discipline wherein the major aim is to explain these very processes which are unique to living organisms and which set them aside from objects of any other kind.

Given that life needs special study, we are faced with a choice between two very different ways to proceed – the 'vitalistic' or 'mechanistic' routes to an understanding of living things. The scientific approach is basically mechanistic – that is, it looks for explanations of life phenomena in terms of understandable mechanisms of causation, not in terms of reference to metaphysical 'vital forces' which defy intellectual analysis. This latter approach is of course vitalism, and in many ways the two are directly opposed. However, this is quite different from saying that a scientific approach to understanding the functioning and immediate origin of living organisms is incompatible with what I shall call an 'enlightened religiousness', which it is certainly not. This issue will be discussed in Chapter 14.

The mechanistic approach to life, with which the vast majority of this book is concerned, is now most often labelled 'biological science'. Biologists are concerned with developing general principles about all aspects of the structure, functioning, interaction and evolution of living organisms – principles which, as we have seen, are separate from, but not in contravention of, the laws of physics and chemistry. Within biology, different disciplines deal with different aspects of order. There are those which deal with particular ordering processes. For example, embryology deals with development, evolutionary biology with evolution, and ecology (or a part of it anyway) with the process of ecological succession. (Succession involves changes in the biota that occur in a given region from its initial colonization up to the establishment of a stable 'climax' community.) Other branches of biology deal with already ordered processes which in a sense maintain the *status quo*. A good example of this is genetics, where the process under investigation (at least in the original form of the subject, now known as transmission genetics) is the mechanism of inheritance, through which, in all species, offspring come to resemble their parents more closely than they do unrelated individuals.

While different branches of biology deal with different problems, it is ultimately necessary that the solutions that geneticists, embryologists and so on come up with are intercompatible. For example, if the mechanism of inheritance proposed as a general solution to the geneticist's problem has features which prohibit evolutionary change, then biologists as a whole are in deep trouble. This situation did in fact happen in the last century. It will be discussed in Chapter 7.

Unfortunately, a biologist today has such a vast literature on his own speciality to keep up with that it is difficult enough for him to stay abreast of the latest findings in his own area, let alone try to relate that area to others, where the people, methods and the nature of the results are all different. There is a clear need for a discipline of Theoretical Biology, devoted to a large extent to making interconnections between developing theories in different specialisms. While a few biologists have made considerable contributions to such a branch of their science – notably the late C. H. Waddington – Theoretical Biology has failed to develop as a concrete entity. This is partly due to the prevailing economic climate, which frowns upon 'pure' research. Partly, however, the fault lies in biology itself. Some biologists denigrate those who deal in speculative generalization, preferring instead the hard facts that derive from a particular study of a certain system in a single species. One aim of this book is to make connections, even where these can only dimly be seen, in the belief that the striving towards synthesis, with all its hazards, is as important an aspect of biology (and indeed of any science) as the more cautious compilation of data on limited systems.

This enthusiasm about generalization is by no means an idiosyncrasy of my own approach to biology. In fact, the striving for formulation of *general* explanations applicable to large classes of phenomena is one of the features that set science aside from other pursuits – such as car maintenance, accountancy and dentistry. Other authors – notably the philosopher of science Karl Popper – have stressed testability as its main distinguishing feature. While this may indeed provide a distinction between science and metaphysics, it is generality of explanation that distinguishes it from non-scientific but practical pursuits such as those listed above. The hypothesis that an engine is running jerkily because of a faulty fuel pump is perfectly testable (by replacing the pump), but the eventual explanation of the problem applies only to that particular car. Unfortunately, in science it is usually the least general and least important questions that are most easily testable – that is, testability and importance/generality are negatively associated. It is relatively easy to test that a population consisting mostly of pale moths evolved into one consisting largely of black moths by natural selection, but the answer to such a test is not particularly general or important. To test whether evolution as a whole is driven by natural selection is vastly more difficult (some would say impossible), but the subject is of course of much greater importance and is much more relevant to the development of a general theory of evolution.

While many questions about life are perfectly acceptable as valid within the framework of biology, including such examples as 'how does life evolve?' and 'how does an adult organism develop from a fertilized egg?', others, such as 'what is the meaning of life?', are not, because they are outside the scope of a mechanistic approach. Actually, with this particular often-asked and seemingly important question, one is tempted to retort with the old clichéd parody of philosophers 'what is the meaning of meaning?'. If in fact the questioner means 'how should a sentient being best live his life?' then it is a question of religion and ethics. If he means 'where did life come from?' then we are back to biology at least for the 'immediate' answer (the origin of life on earth).

Unless we do phrase it in one of these more specific ways, the question 'what is the meaning of life?' is so obscure that '42' may well be as good an answer as any.* In the pages that follow, I hope to deal with more concrete questions and with more satisfactory, if sometimes tentative, answers.

*See *Hitch-hikers' Guide to the Galaxy* by D. Adams.

2 WHAT IS LIFE AND WHAT ARE THEORIES?

Most people, if asked to give an example of life or an example of a theory, would have no difficulty whatever. Blackbirds and pine-trees are obviously alive while rocks and clouds are not. Natural selection and relativity are clearly scientific theories and are often referred to as such, while the weather forecast for a particular day would hardly be regarded as having a similar status. However, being able to give an example of some category of phenomena is not the same as being able to define it, and if asked to define either life or a scientific theory, most people would find the task rather more taxing. While for most terms used in this book a general idea of what they mean is sufficient (as with order in the previous chapter), it makes sense, at the outset of an investigation of 'theories of life', that we should pause to consider in detail what is meant by our two most central terms; otherwise you, the reader, and I, the author, may have different mental pictures of what lies behind the convenient verbal labels of 'theory' and 'life' and the labels will serve to confuse rather than to inform. I do not pretend that we can reach universally acceptable, cast-iron definitions, but I do think that we can refine our ideas greatly, and reach a good degree of consensus, by briefly considering this problem.

'Life' is the easier of the two to define, so I shall start with that. There are two very different ways to proceed in this case, which can be thought of as the chemical and the evolutionary approach. The chemical approach seeks to define living organisms in terms of the substances they contain and the properties of those substances. The evolutionary approach, on the other hand, disregards the chemical make-up of living organisms, and attempts to define them on the basis of whether or not they are subject to natural selection. Both of these approaches have their different problems, yet they yield essentially the same answer, which is encouraging. I shall now examine each approach in turn.

As regards their basic chemistry, particularly that relating to their genes, all entities we normally regard as life-forms are remarkably similar. Organisms as diverse as man, bacteria, viruses, frogs and ferns all have genes composed of a substance called nucleic acid. In all of these, except some viruses, the type of nucleic acid found is the 'deoxyribose' type – hence DNA. In viruses, this is sometimes replaced by RNA – ribonucleic acid. In all cases the genes are responsible for the manufacture of proteins, which contribute to both the structure of the organism – such as its various membranes – and its chemical functioning – in which the protein catalysts we refer to as enzymes are of paramount importance. As well as producing proteins, the genes must be able to replicate – that is, to copy themselves and produce more identical sets of genes. This is necessary for two reasons. First, so that, as an organism grows, each of the increasingly large number of cells has its own set of genes, and second, so that the organisms can ultimately reproduce.

Using these criteria, then, we can define a living organism as any entity containing nucleic acid capable of replicating itself and of making proteins. For the most part, classifying objects with this definition corresponds with the classification that we would make on the basis of intuition and everyday experience. For example, the definition tells us that blackbirds and pine-trees are alive while rocks and clouds are not, which was where we started. A difficulty arises when we consider the 'primitive' forms of life, particularly viruses. These are indeed living organisms according to our definition, provided that we do not insist that our life-forms must be capable of gene-replication and protein-making without outside assistance. If we were so to demand, viruses would cease to be alive, since they parasitize the cells of animals, plants and bacteria, using the genetic machinery of their host to produce more virus genes and proteins. So viruses constitute a 'grey area' on the border of life, falling to one side or other of the dividing line depending on precisely how we phrase our definition. Indeed, as we shall see, this area has become even greyer of late, with the discovery of other virus-like objects, notably 'retroviruses', which are in a sense midway between viruses and genes.

Another difficulty arises if we want our definition of life to apply on planets other than the earth. We are still totally ignorant about whether life exists elsewhere in the universe and if so what form it takes. The only thing that we do know with reasonable certainty is that any other forms of life that do exist are extremely distant. Searches within the bounds of our own solar system – notably the NASA Viking lander's experiments on Mars – have turned out to be negative. But the solar system is only an infinitesimally small fraction of the universe, and planets containing life of some sort may indeed exist in other solar systems, or even further afield in galaxies outside our own Milky Way. If life does indeed exist in any such places, would we recognize it as such? At the visual and intuitive level, the answer to this question would depend almost entirely on physical appearance. A 'humanoid' as seen in various sci-fi films would be instantly distinct from his planetary background, but a rock-like organism would be no more intuitively recognizable to us as life than an earthly stone cactus in a bed of pebbles. Suppose, then, we are able to make chemical analyses and apply our chemical definition of life: would this help us to separate life from non-life on a distant planet?

The answer to this question depends on whether it is possible to have non-carbon-based life, or whether, given carbon as a base, it is possible to have non-nucleic acid/protein-based life. If either of these is possible, our definition ceases to be useful. Thus we should ask ourselves: what general features do we think of as life-like regardless of the chemical constitution of the objects displaying them? Perhaps these features can be built into a more broadly applicable definition. We must start, of course, by ruling out features specific to certain kinds of life. For example, animals are usually mobile, but plants are not. However, three features are characteristic of all terrestrial life-forms and are likely to be displayed also by alien organisms if indeed these exist. The three features are: reproduction, variation and inheritance. These could be used to form the basis of an alternative definition, which might go something like this: Any object capable of reproducing itself and conferring at least some of its own traits on its offspring, but which is not identical to any of its offspring (or indeed to any other object at all) constitutes a living organism. As it happens, these are the three characteristics required for natural selection to operate on a group of entities, so we have now arrived at the 'evolutionary' definition of life.

We have seen that an advantage of this definition over the chemical one is that it is potentially applicable to non-carbon-based living systems which may one day be found on distant planets. Does it have any disadvantages? There is one to which we should briefly turn our attention, which may be referred to as 'circularity'. This really lies in the way I have arrived at the evolutionary definition. The route I took was via the inadequacies of the chemical approach and the need to have a definition that told us, for example, that a non-carbon-based humanoid who obviously looked alive was indeed a life-form. But this presupposes that we can already recognize a living being! There is no real problem here, but we need to recognize that definitions are merely a way of trying to encapsulate the key features that we recognize in something – in this case life – and to make them explicit. In a way, defining something such as life is just an exercise in refining our ideas. The form the refinement takes is one of getting rid of features characterizing some of a group of objects and concentrating on those applicable to all of them – but to no others outside that group. Thus, with the group we call life, we dispose of four-leggedness because it applies only to a small sub-group of organisms but also three-dimensionality, as it applies not only to all life-forms but also to rocks, clouds and other non-living things. By concentrating on nucleic acid and protein, or on reproduction, variation and inheritance, we are isolating those sets of characteristics that we think apply to all life-forms but not to any non-living objects – these features are what a mathematician would call the 'necessary and sufficient conditions' for life. Once a definition is good enough, we should allow it to supersede our intuition when it comes to classifying future discoveries; arguably, the evolutionary definition of life is indeed good enough.

We now come to the more difficult problem of arriving at some common view of what constitutes a scientific theory. Doubtless, philosophers of science could argue for days on end over this topic, but the purpose here is hardly the scholarly analysis of the term 'theory' but rather the achievement of some sort of working definition sufficient to avoid confusion. In order to proceed in this direction, it will help if we contrast a 'theory' with the related concepts of hypothesis and law. This is not an exhaustive list of theory-like things, but it is enough for our present purposes. Some other related terms are not very useful – for example 'thesis', which nowadays more often means a weighty volume presented for a Ph.D. than the argument it contains!

One of the main distinctions between these different types of 'scientific idea' is the degree of certainty or uncertainty currently attached to it. A hypothesis is the least certain and is essentially a tentative answer to a particular question, often put forward so that it can be tested experimentally. For example, a common observation in biology is that within a particular species – whether of flies or snails or plants – individuals grow to a larger size when the overall population density is low than they do when conditions are more crowded. This raises the question: what causes retarded growth in crowded conditions? There are several possible answers, two of the most obvious being lack of food or build-up of toxic waste products. A scientist studying this phenomenon might – for whatever reason – favour the first of these two alternatives. In this case he would probably put forward a hypothesis something like: lack of food in high-density situations causes decreased growth. This is of course just one of the possible answers put forward as an untested assertion. To test the assertion, or hypothesis, our scientist might design an experiment where food is no longer limiting in high-density cultures of the organisms he is studying. Thus he is separating the two potential answers to his question. If growth is still retarded in his experiment, we presume that his limited-food hypothesis was wrong.

At the other end of the spectrum of certainty/uncertainty is the scientific law. These sometimes come individually but more often in sets. An example of the latter is provided by the Laws of Thermodynamics, which are of central importance in physics, and indeed in science as a whole. (The First Law of Thermodynamics is better known by its alternative title, the Law of Conservation of Energy.) In biology, there are relatively few laws, compared with the physical sciences, but we do have Mendel's laws in genetics. Basically, a law is a universally accepted generalization. Usually, universal acceptance can be achieved in science only when very precise testing is both possible to conduct and positive in outcome when conducted; and precise testing is possible only when the assertion to be tested takes a quantitative form. This is why laws are rare in biology, and also why the best example comes from genetics, which is a quantitative branch of biology, rather than – say – anatomy, which is much more descriptive.

Using these two reference points – a hypothesis, which remains to be tested, and a law, which has been rigorously tested and has borne up to its tests – where does a theory fit in? The answer is, unfortunately, that it can fit in almost anywhere, and that its meaning is much more difficult to pin down than that of a hypothesis or law. On the one hand, theory can mean hypothesis, as it does in the phrase 'it's only a theory', meaning either that an idea has not yet been adequately tested or that it has not been tested at all. On the other hand, in the Theory of Evolution we are referring to a process – evolution – which is almost universally accepted within the present-day scientific community and thus has almost the degree of certainty associated with a law. In this book, I shall use 'theory' in this latter sense.

There are two other dimensions in addition to certainty/uncertainty which are helpful in distinguishing between hypotheses, theories and laws. One that has already been briefly alluded to is the degree of 'quantitativeness'. This is most useful in distinguishing laws (such as Mendel's) from otherwise law-like theories (such as Darwin's). The other important dimension is breadth, which helps to separate hypotheses from laws and theories, because the former are often quite specific while the latter two are usually broad. Thus both Darwin and Mendel were ultimately making statements broadly applicable to all characteristics of all species in the living world, whereas the hypothesis that, for example, high-fat diets cause heart-attacks in man is of relatively narrow applicability and would not, if proved to be true, turn into a law or theory.

Before leaving this area of the relationship between hypotheses, theories and laws let us consider further what happens to hypotheses over time, as they progress from untested to tested. One thing that is obvious is that, if our testing is competent to deliver a yes/no answer, a hypothesis cannot remain as such after testing. If it turns out to be false, it is simply a 'refuted hypothesis' and disappears from the face of science. Suppose, however, that it turns out to be true: what then? Basically, it turns into either a theory or a law or, if it is too specific and limited to be either, a fact. A particular diet known to cause a heart condition in humans is a fact. That humans, generally speaking, have five fingers per hand is a fact, as is the occurrence of a particular group of four nitrogenous bases in DNA. So far, all seems in order: most (narrow) hypotheses, if proven, turn into facts. A few broad hypotheses turn into theories or laws depending on how quantitative they are. The only complication is that the apparent breadth of an idea may change as it gets tested. If Mendel's view of heredity had turned out to be true only of garden peas, it would have been an isolated fact. Since it turned out to be *generally* true, we now have Mendel's laws.

I mentioned earlier the possibility of a hypothesis being 'proved to be true'. According to the philosopher of science Karl Popper, this is not possible. In Popper's view of science, it is possible only to *dis*prove a theory, not to prove it. Whether or not this is a sensible view of science depends on the exact form in which hypotheses are put forward. If we state that 'all human hands have five fingers' and conduct a survey of 1000 hands all of which do turn out to have five fingers, we have not proved the hypothesis, because there might be other human hands, not monitored in our survey, that have six fingers. These do in fact exist; the phenomenon is called polydactyly and is caused by an abnormal gene. However, had one person with polydactyly been included in our 1000-hand survey, we would have conclusively disproved our hypothesis in the form in which we initially stated it.

The problem with Popper's approach, at least in biology, is that hypotheses are not usually put forward as tentative *universal* truths, but rather as tentative *general* ones. This is the case regardless of whether the hypothesis relates to a specific or very broad issue. Thus we are content to know that *most* human hands have five fingers; also that *most* species have two copies of each gene, as Mendel proposed (a few have only one copy). If we acknowledge that biology works in this non-universal manner, it follows that Popper's approach must be employed with caution.

One final point about the nature of scientific theories needs to be made – a point which may well be more important than any of the preceding ones, even though attention is rarely drawn to it. This is that all scientific theories arise in pairs or, occasionally, in groups. In other words, we usually have, at least for a time, *opposing* theories – alternative and mutually exclusive views of the living world. We see several such clashes between opposing world-views in recent and current biology: evolution versus creation, blending versus particulate inheritance, and regulated versus unregulated populations, to name just a few. These and other clashes will be recounted in subsequent chapters.

It is interesting that certain major fields of biology appear to be without general theories. Among these is anatomy, which deals with the gross structure of organisms.* Also, biochemists have not come up with opposing theories of metabolism. Why are such disparate fields apparently devoid of general theoretical content? One can only guess at the answer to this question. My guess would be that the reasons are different in the two examples given. Anatomy is a largely descriptive science; theories are about how things work. If you are studying the form that something takes rather than the way it works, it is difficult to come up with opposing theories – you merely describe how things are. There may be an element of this in metabolic biochemistry also. The working out of the metabolic network – the pathways of chemical reactions which power living organisms – was, despite the experimental nature of the techniques, a largely descriptive task. However, there may be another reason why there are no opposing theories of metabolism. Theories often arise when scientists think long and hard about the nature of a process and come up with possible explanations before any experimental 'handle' is found with which to begin an exploration of the facts. The success of biochemical techniques in rapidly revealing the main metabolic pathways may have prevented opposing theories from arising by, essentially, providing the facts first.

*In *comparative* anatomy, however, homology constitutes a general theory.

It should now be clear exactly what this book is about: broadly applicable, testable and generally accepted explanations that apply to some major aspect of the functioning, evolution and interactions of living organisms; and threats to these explanations from non-scientific groups to whom the testability of a theory is unimportant. Anyone with a biological background should now proceed direct to Chapter 4; other readers are advised to read Chapter 3, though it can alternatively be skipped and used only for reference when an unfamiliar term crops up later on in the book.

3 A DETACHABLE BASIC BIOLOGY COURSE

If you browse through the appropriate shelves in the library of an academic institution, you will find plenty of books on introductory biology. Often, these run to 200 or 300 pages. In this book, however, I can devote only a chapter to the basics, before we proceed to take a deeper look at the most important biological theories. Now, while the author of a 300-page introductory text may be guilty of some unnecessary verbosity, it is doubtful if the excision of this would reduce his book by 270 pages or so! Thus it is obvious that I cannot hope to cover here all of the 'basic biology' that there is to know.

An author faced with too little space to present too much material has two choices: he can be either superficial or selective. I have taken the second option, both because a superficial treatment of lots of bits and pieces is of little use to anybody, and because selectivity is harmless, perhaps even helpful, when we need to know only the basic biology behind certain major theories. The one way in which selectivity might be dangerous is if it was not recognizable as such, leading the reader to think that he has covered the whole range of biological science when he has not. That is why I am making this point now. Readers with a fascination for the anatomy of reptiles, the courting behaviour of a praying mantis or the ecology of rattlesnakes in American deserts will have to look elsewhere to satisfy their curiosity. What this chapter is concerned with is the basic biology necessary for an understanding of the theories discussed in the rest of the book, and nothing else. Even this is a tall order.

Levels of life

Biology used to be most often divided up according to what kind of organism was being studied. Thus zoologists studied animals,

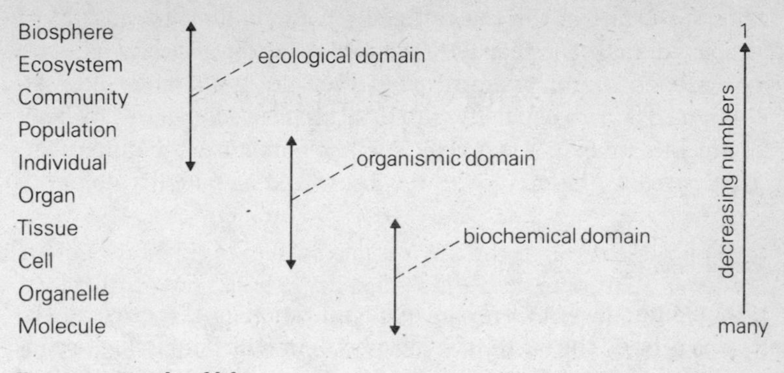

Figure 1. Levels of life.

botanists plants, and microbiologists the various 'bugs' with which the layman is much less well acquainted. While these divisions have not entirely disappeared, it is commoner, nowadays, to divide biology up according to the level of study, which can be anything from molecular to ecological. The levels of organization exhibited by living systems are given in Figure 1, which also shows the range of levels which some particular biological sciences focus upon. These disciplines (ecology and biochemistry, for instance) are sometimes seen as resulting from a 'horizontal' carve-up of the living world, in contrast with zoology, botany and microbiology, which appear as a result of a 'vertical' split. These phrases connect with a mental image of a table in which the row headings are the levels of life given in Figure 1 and the column headings are animals, plants and micro-organisms.

One advantage of this mental picture is that it shows that the work of a particular biologist falls into a particular 'box' (or sometimes several boxes) and that the biologist concerned can thus be described in one of two ways. For example, someone studying an animal population can call himself a zoologist or an ecologist. Equally, someone studying the functioning of a plant at the molecular level can think of himself as a botanist or a biochemist. As I have said, the horizontal split into 'levels of life' is predominant in modern biology, and the structure of the rest of this chapter reflects this. Basically, each of the following sections (except the last two) deals with a particular level of study. I have chosen to start at the top rather than the bottom because most people are more familiar with ecosystems (examples of which are ponds, woods and swamps) than they are with molecules. Although it is rarely explicitly stated, all sensible education – by book or any other means – proceeds from the familiar to the unfamiliar. To proceed in the opposite direction would be patently daft.

Basic ecology

The highest level of ecological organization is the circular envelope around the earth (the biosphere) in which all living organisms are found. The biosphere is of course far too large and complex to study as a functional unit, so at the very least it must be broken down into the smaller units that we call ecosystems for the purpose of scientific study. An ecosystem is the total biota present in a particular area, together with abiotic substances such as soil, air and water. Sometimes the boundaries of an ecosystem are very precise. For example, a single stagnant pond in the middle of a large forest is a particularly clear-cut ecosystem. More often, the boundaries of an ecosystem are more difficult to pin down, and are set arbitrarily for the purpose of study. For example, if we wanted to investigate ecological processes occurring in the vast coniferous forest stretching over much of Canada (the 'boreal forest'), we might mark off a one-acre 'plot' as our 'study-ecosystem', even though there is no physical separation of this area from the others around it.

A community is simply the biotic component of an ecosystem. If we were to instantly vaporize our stagnant pond, leaving a bewildered array of water snails, pondweed, insect larvae, bacteria and perhaps a couple of ducks, this collection of organisms, considered in isolation from its physical surroundings, of which we have conveniently disposed, constitutes a community.

Of course, a community is not just an orderless heap of organisms. Rather, it is structured in all sorts of ways. One of these is the structure that we refer to as a food web. Suppose our pond contains, among other things, four species of algae, two species of pond snails and a duck. Each snail species eats a range of algal species, and the snails in turn are eaten by the duck (see Figure 2).

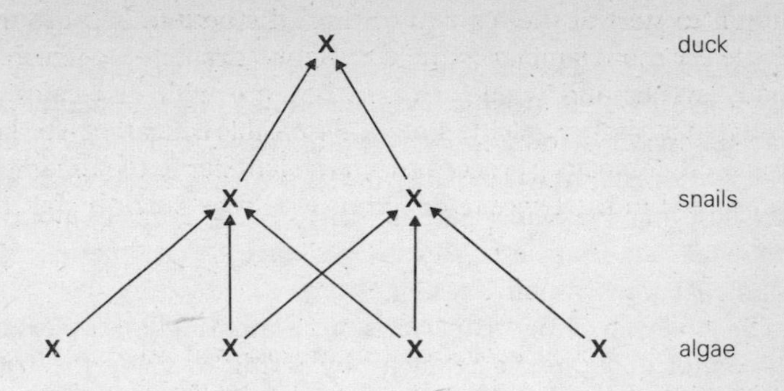

Figure 2. Food web of a hypothetical pond. (× = one species. The arrows indicate energy flow.)

This pattern of interconnection (which goes 4 → 2 → 1 in terms of species numbers) is a food web or, more correctly, *part* of the overall food web of the whole ecosystem, the pond. The sort of web I have just described, where the number of species declines as we go 'up' through it, is quite common, but not universal. For example, many species of insect are supported by a single species of tree in the boreal forest, so here, as we come 'up' from plant to herbivore, the number of species increases. (If we go further up, to birds, it decreases again.)

I have begun to talk about going 'upwards' in a food web, which is a way of picturing the direction of energy flow from food-organism to consumer-organism. If we think of plants as the bottom level of the food web, then the next level up is herbivores, and this is followed by one or more layer of carnivores, ending with the so-called 'top carnivores', examples of which, in different ecosystems, are whales, lions and man. Of course, even top carnivores are consumed when they die. This, and consumption of dead organisms from all other levels, is achieved by scavengers such as vultures, and by an inconspicuous but important part of the food web – the 'decomposers', which are mostly bacteria and other micro-organisms.

Sometimes the various levels – plant, herbivore and so on – are referred to as trophic levels. (Trophic simply means 'feeding'.) This is a useful label in some ways, but we have to bear in mind an important problem besetting it, namely that many kinds of organism do not fit neatly into one or other level. We could say that a plant was in trophic level 1 without much difficulty, and that an insect feeding on the plant was in trophic level 2 (= herbivores). But what of a bird that sometimes eats seeds from the plant and sometimes eats the insect? And what of a bird-of-prey that eats that bird? Clearly, we have to use the idea of a trophic level with some caution.

The food web of any real ecosystem is amazingly complex, and we cannot make much headway by trying to study the whole thing. Consequently, ecologists often focus on a particular species – one of the species of water snail in our stagnant pond, for example – and concentrate on that. They can then examine the links with its food-species (the algae) or its predator-species (the duck) in considerable detail. (Note that terms like food-species, consumer and predator are ways of referring to what is below or above a particular species in the food web, without implying anything about what 'level' that species is at – in contrast to terms like herbivore and carnivore.)

Concentrating our efforts on a particular species in a particular ecosystem brings us to the study of populations, since that is precisely what a population is. To some extent, the degree to which a population is clearly delimited corresponds to how clear-cut is the ecosystem in which it finds itself. Our species of snail in its stagnant pond is well off in this respect. Yet these two kinds of delineation do not always go together. If our pond snail species inhabits a large lake, there may be concentrations of it in various bays of the lake which have relatively little interaction with each other. Whether or not we should call these separate populations is not immediately obvious.

Within even a small ecosystem, any particular population occupies a restricted physical position. Birds often live in trees, some insects in crevices in rocks, and moles underground, for example. We refer to these restricted physical places as the microhabitats of the populations concerned. In contrast with this, the *niche* of a population is its food habits, or, according to some authors, its place in the food web, which means its relationship with predators as well as with food-organisms. The emphasis is usually on food, though. Niches are important because species whose niches overlap may come into competition with each other, while those with non-overlapping niches cannot do so.

Populations feature quite strongly in this book, while communities and ecosystems receive very little mention. The reason for this contrast is simply that there are well-established theories relating to populations, while the theories of community and ecosystem ecology largely remain to be worked out.

Basic organismic biology

The ecologist investigating the growth or regulation of an animal population is essentially treating the organisms that make up the population as 'units' or 'black boxes'. He needs to know that they consume food, develop, breed and die, but he does not specifically investigate how, for example, each 'unit' of his population grows from a fertilized egg into an adult animal. Questions of that kind are beneath the ecological sphere, and fall into a heterogeneous zone between ecology and biochemistry that can be labelled 'organismic biology'.

The study of the gross structure and function of adult organisms, which was at the centre of biology in previous centuries, is not active today, and is certainly not a sphere which has given rise to general theories of the sort discussed later in this book. So if you want to know about skeletons, or the structure of the gut, you will have to consult the type of general introductory biology text I mentioned earlier. From our point of view as aspiring theoreticians, what is of more interest than a static view of an organism at its adult stage is a dynamic view of it as something that develops, breeds and ages. In other words, we are particularly interested in life-cycles.

Of the three life-cycle processes just mentioned – development, reproduction and ageing – I shall concentrate on the first two. How and why organisms age is currently receiving much study, but as yet we have no accepted general theory of ageing (nor a cure for it!). On the other hand, we do have a theory of how organisms inherit their characteristics via the reproductive process; and as for development, we at least have a theory about one of its major components, and we may well be closer to a general theory of development than some biologists think.

Since we all know that eggs precede chickens, I shall treat reproduction first and then move on to development. (A more balanced discussion of the 'chicken and egg' puzzle will be given later.) There are two basic types of reproduction – sexual and asexual. Sexual reproduction, in which male and female parents both contribute a germ-cell, or gamete (sperm and egg in animals, pollen and ovum in plants), to the formation of the new organism, is the prevalent form of reproduction among higher organisms, particularly animals. Asexual reproduction is found more in plants and in unicellular creatures. In the latter group, two important types of asexual reproduction are (a) budding, in which a small daughter-cell buds off from its parental cell, and (b) fission, where the parent-cell splits into two identical daughter-cells. In asexual reproduction, each offspring receives all the parent's genes, and is essentially a replicate of the parent. A group of such replicate cells comprises a clone. In sexual reproduction, the male and female gametes fuse together to form a single cell, the zygote, which will develop into a new organism. In this case, both parents contribute half their genes to each offspring but, unlike the germ-cells containing them, the paternal and maternal gene sets do not fuse together. Rather, they form a sort of temporary combination, according to the rules of inheritance worked out by Mendel.

In plants and animals, the adult form is composed of many cells. The starting point, the zygote, is but a single cell. Somehow we have to get from one to many, and not only this, but the many must be organized into a very strict spatial pattern. The 'many' are also ultimately very different in type – blood cells, nerve cells and so on. The whole process – growth, diversification of cell type and production of overall spatial pattern – is simply called development. Within this overall process, the production of different specialized cell types is referred to as cell differentiation.

The early study of development was largely descriptive, and, as a result of this study, we know a lot about the sequence of stages through which a developing organism goes. For example, in vertebrates, the zygote is followed by a solid ball of cells called the morula, then by a hollow 'sphere' of cells called the blastula, and then by a much more complex arrangement called the gastrula. Many later embryonic stages intercede between this and the recognizable 'young adult', for instance the mammal just after birth. Some kinds of organism, such as insects and frogs, make life even more difficult by having a multi-stage life-cycle in which at least two largely separate developmental processes occur.

The descriptive study of development, for which I shall use the label 'embryology', has now been largely superseded by a more analytical approach, whose practitioners tend to refer to themselves as developmental biologists, rather than embryologists, to emphasize their more experimental approach. (Actually, some of the work of the earlier embryologists was experimental also.) Developmental biologists are thus concerned with the question of *how* cells differentiate and spatial patterns of cells form, rather than in what exact sequence of steps these things occur in any particular organism.

One of the many adult cell types to which the process of cell differentiation leads is the germ-cell or gamete. We thus come full circle, because at the *end* of development we have got, among other things, the cells necessary to *start* development. Perhaps the chicken precedes the egg after all.

Basic cell biology

I have jumped ahead in the previous section and referred to cells before explaining what they are. I did this (a) because it was impossible not to, and (b) because I suspect that even the most abiological (as opposed to abiotic!) reader will have some rough idea as to what cells are, and will be aware that nearly all organisms are made of them. Actually, describing an organism as being made of cells by-passes a couple of other levels of organization. Organisms are made up of organs (brain, heart, etc.); organs often contain several tissues (hearts, for example, contain much muscle tissue, some connective tissue and some nerve tissue); and tissues are made up of cells. A sheet of muscle tissue, for example, is just a great mass of interconnected muscle cells. Describing the overall animal (or plant) as being made up of cells is not inac-

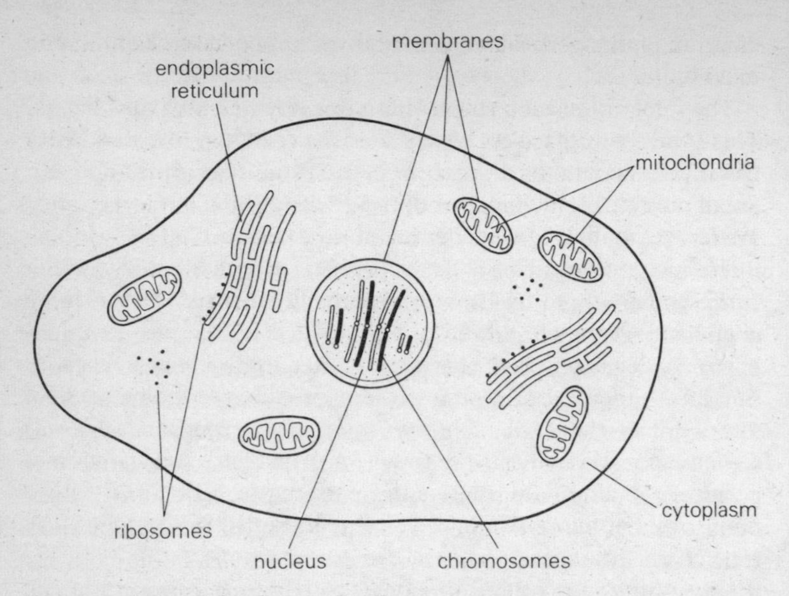

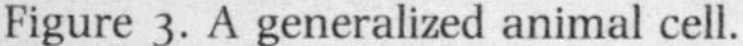
Figure 3. A generalized animal cell.

curate, but we should remember that we are leaping over a couple of levels of organization.

There are at least four different ways we could discuss cells. First, there is the question of what the hypothetical 'generalized cell' looks like from a static viewpoint. Then there is the dynamic approach – adding the temporal dimension and looking at a cell's long-term behaviour, particularly the process of cell division by which one cell gives rise to another. The third way we can look at cells is to describe the diversity of different cell types that exists. Finally, we can ask how this diversity arises.

We shall look at the first two of these now. How cells differentiate is the subject of a later chapter, and I shall give sufficient descriptive details there of the actual diversity of cell types that exists, so that we know what there is to be explained in terms of differentiation.

On, then, to the 'generalized cell', by which I mean a cell which has all the main attributes which characterize most cell types, but none of the specialist attributes (muscle fibres, chloroplasts, etc.) that are found only in one or a few cell types. Figure 3 is a diagram of this generalized cell, and we now need to examine its main features.

The integrity of each cell is maintained by its outer membrane. This is present regardless of whether the cell is on its own (like a blood cell) or part of a great mass of tissue (like a muscle cell). Small molecules can pass through the membrane, but larger ones are unable to do so. Inside the membrane, the bulk of the cellular interior is composed of a heterogeneous substance we call the cytoplasm (literally 'cell material') in which many of the cell's chemical reactions occur. Suspended in the cytoplasm are various membrane-bound organelles. Most important is the central nucleus, which contains the chromosomes and the genetic information they embody. Also relevant here are the mitochondria, sometimes described as the powerhouses of the cell, since important energy-releasing reactions go on within them, and a series of membrane-bound channels collectively called the endoplasmic reticulum (ER). Small bodies called ribosomes are found both on the ER and loose in the cytoplasm. These are the sites of protein synthesis in the cell.

On a short-term, minute-to-minute basis, the cell is simply ticking over chemically, keeping itself supplied with energy, making proteins from the genetic information contained in its chromosomes, and letting those proteins go about their business. In the longer term, however, it must divide to produce daughter-cells. This is necessary because the organism of which the cell is a part lives far longer than any of the cells within it. One of the epithelial cells in your stomach may only live for a day (partly because it is continually subject to hydrochloric acid in the stomach itself). One of your blood cells, on the other hand, may last for a hundred days. In either case, the cell's lifetime is a long way short of seventy years.

If you were simply to put a partition through the middle of the cell shown in Figure 3, would that cause any problems? Actually, for most purposes it would serve as a reasonable way to divide the cell. A vertical partition would leave right-hand and left-hand cells with ribosomes, ER and mitochondria, as well as about half of the original cytoplasm. Given the ability to grow, the daughter-cells will be fine. If the partition is made of cell membrane, then when it is complete, it leaves two fully membrane-bound cells. The only problem, as will be apparent by now, is that the nucleus is going to get in the way.

This problem cannot be solved simply by driving our partition straight through the nucleus. This would leave each daughter-cell not only with a damaged nucleus but also with only half of the genetic information it requires. Instead, the chromosomes must produce a duplicate set of themselves so that each daughter-cell can have one set. This in fact is the first thing that occurs in cell division (along with disintegration of the nuclear membrane). It is followed by migration of the two sets of chromosomes to opposite ends of the cell, after which new nuclear membranes form around each set. It is only then, *after* we have got two daughter nuclei, that the cell is partitioned by growth of, essentially, a vertical partition of cell membrane.

This whole process of chromosomal duplication and cell division, which occurs millions of times in the development and maintenance of a human body, is called mitosis. It is shown, in simplified form, in Figure 4. The only situation where cell division significantly deviates from this basic pattern – in man or any other organism – is in the formation of the germ-cells that go on to form the next generation. The *reproductive* kind of cell division, called meiosis, differs from mitosis for the following reason. In mitosis there is one doubling of chromosomal material, and one halving of it, so we end up back where we started, as we should do if the daughter-cells are to be like their parents. In meiosis, we are producing a germ-cell which will fuse eventually with a germ-cell from the alternative sex, thus causing a second doubling of the genetic material. To compensate for this, we need a second halving. Meiosis is thus more complex than mitosis. It is also much rarer and more intimately associated with reproduction and, thus, with the genetic link from one generation to the next. Because of this, we shall postpone examination of meiosis until the 'basic genetics' section, which comes after the following one.

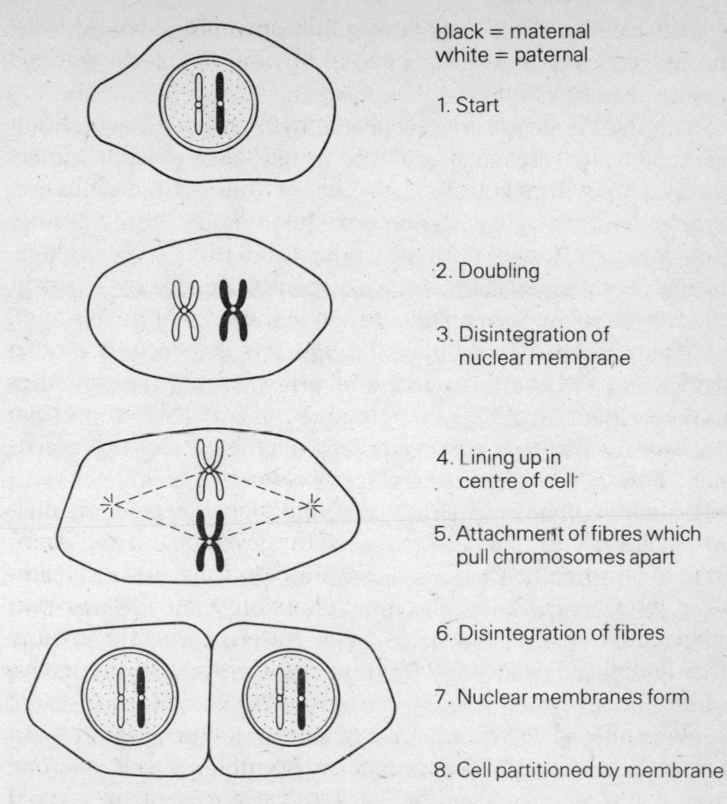

Figure 4. Mitosis.

Basic molecular biology

There is a subtle difference between 'biochemistry' and 'molecular biology', although there is no reason to suspect it from the terms themselves. For no particularly good reason, molecular biology tends to refer to studies of the genes themselves (DNA molecules) and those molecules that are closely associated with them (RNA and protein). Before the advent of molecular biology, people calling themselves biochemists were investigating *all* biologically important molecules. Now, however, 'biochemistry' seems almost to imply a concentration on those sorts of molecules that are *not* central to the molecular biologist, such as carbohydrates and fats. The distinction is not quite as clear-cut as this, because what *aspect* of a molecule is being studied is also important. If you are studying how a gene makes a protein, you are a molecular biologist; if you are studying how that protein catalyses a chemical reaction, you would probably consider yourself a biochemist. Since I have little to say in this book about molecules not directly related to DNA, what follows is molecular biology. In fact, molecular biology so defined is essentially molecular genetics. So, if you like, this section is part of the next one.

Within molecular genetics, I want to concentrate on the structure of DNA, the structure of proteins, and how the two connect – that is to say how a particular piece of DNA comprising a gene gives rise to a particular protein. I shall also say a little about how DNA replicates itself and about what proteins do. I shall not discuss how genes are controlled, but there is something on this later.

We start, then, with the substance of the genes themselves – DNA. As everyone knows by now, the basic shape this molecule takes is that of a double helix – two helices coiled round each other and linked chemically together. As it turns out, it is not the helices that are genetically important, but the links between them. Consequently, it will do no harm if we mentally straighten out the helices and treat them as two straight lines linked together. The DIY fanatic will easily recognize this structure – it's a ladder!

The two uprights of the ladder (the helices) are made of alternating sugar and phosphate molecules. I mention this out of habit – you don't really need to know it. What is important is that the uprights are linked together in a very specific way. Each upright has a series of molecules projecting from one side. These are known collectively as nitrogenous bases and individually as adenine, guanine, cytosine and thymine. Since no one likes names like this, they're usually just called A, G, C and T. Because of their chemical nature, A on one upright always pairs with T on the other; similarly, G and C pair together. Thus a short stretch out of a long DNA molecule lying on its side looks something like this:

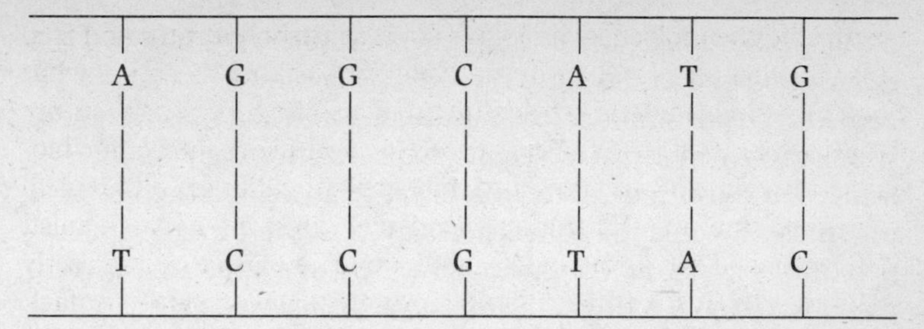

The reason A, G, C and T are important is that it is the sequence of these along one of the two strands (the 'sense' strand) that tells the protein-synthesizing apparatus what protein to make. But before going into the details of this, let's turn and have a quick look at the structure of proteins.

Basically, a protein is just a long string of smaller molecules called amino acids. There are twenty *different* amino acids, and a particular protein will be represented by a particular sequence of amino acids. Most proteins are between fifty and 500 amino acids long. Actually, the amino-acid sequence is just the 'primary' level of structure of a protein, just as the ladder represents the primary structure of DNA. Between amino-acid string and functional protein, and between ladder and chromosome, several levels of folding and subunit-binding occur. These are not really relevant here; the main DNA-protein link is at the level of primary structure.

The mechanism of protein synthesis is remarkably simple, at least in terms of the way the message is coded. DNA bases are read off the sense strand in groups of three ('triplet codes'), and copied into their RNA mirror-image (for instance TCC on DNA into AGG on RNA). Each RNA triplet is interpreted by the protein-building machinery as a code for a particular amino acid. For example, AGG on RNA is the code for the amino acid called arginine. Thus a complete gene, between punctuation signals, is read off into a complementary RNA molecule, which in turn is read off into an amino-acid sequence, which is linked up enzymatically into a protein. If we started with a gene of 600 bases, we end up with protein of 200 amino acids.

A complication that has recently been discovered is that many genes in higher organisms have 'intervening sequences' (or introns) that are *not* turned into protein. Rather, the bits of RNA that correspond to these are removed before the RNA leaves the nucleus. This is a minor complication, but it does lead into the issue of *where* all these molecular happenings are going on. Basically, the copying of DNA into RNA (transcription) goes on in the nucleus, while the conversion of RNA into protein (translation) takes place at the ribosomes. At some time in between, therefore, the RNA leaves the nucleus. Since it is too big to diffuse through the nuclear membrane, it exits via convenient holes in the membrane known as nuclear pores.

To the reader with little prior biological background, the process of protein synthesis described above may seem like just one out of a myriad of chemical reactions going on in the body; and it may seem odd that we focus so much upon it. Let me try to explain why we do adopt such a focus, despite the fact that there are indeed many other sorts of biochemical reaction. As we have seen, in all organisms except for some viruses, DNA is the genetic material. So *all* genetic information resides in DNA. The nucleus is the 'home' of most of the DNA. It is true that there is a little DNA in some organelles, such as the mitochondrion, and that the nuclear chromosomes contain some proteins as well as DNA. Nevertheless, a basic idea of genetic information being carried by DNA which forms the chromosomes that we see in the nucleus is not too far out. In short, 'DNA is information'. On the other side of the coin, we can say that 'protein is actuality'. At whatever level we view the body (of a human or other organism), proteins are a major component. At the cell level, proteins are a major component of all membranes; at the tissue level, they are a major component of muscle; at the level of external appearance, your hair is essentially a mass of protein. As well as being a major component of structure, they are the main agent of function. Many proteins are enzymes, and these catalyse the body's chemical reactions. So the body as a whole is largely made of and run by proteins. The mechanism of protein synthesis, therefore, is in many ways the key process of life at the molecular level, as it represents the transformation of information (DNA) into actuality (proteins, and hence body). As we have seen, RNA is the 'message' which passes from one of these to the other.

Just as organisms have life-cycles, so the cell is characterized by a 'cell-cycle'. The process of protein synthesis goes on during a phase between consecutive cell divisions (mitoses) that is called interphase. The cell-cycle, in simplified form, is an alternating series – mitosis/interphase/mitosis/interphase – going on indefinitely. The relative length of time spent in interphase and mitosis varies a lot depending on what cell you are dealing with, but in all cells the association of protein synthesis with interphase holds true. During mitosis, on the other hand, the chromosomes (and the genes they carry) become inert and are essentially just packages that are shuffled about. However, immediately before mitosis (or just as it begins, depending on exactly which stage you designate as the start of it) the DNA duplicates. We have already seen this at the level of the gross chromosome, and we know that this duplication is necessary if both daughter-cells are to have a complete chromosomal complement. But we have not yet seen how this duplication is achieved at the molecular level.

Actually, the process is quite closely related to the transcription stage of protein synthesis in that it involves pairing between the bases A and T, and between C and G. To look at DNA replication, as it is called, at a simple level, I shall again treat the molecule as a ladder rather than a double helix. What happens is not unlike what would happen if our DIY enthusiast got mad with his ladder and pulled the two uprights apart at one end, shearing the rungs in the middle. In the case of DNA, it is an enzyme that does the pulling apart, and the 'rungs' always split right at the middle because the link between a pair of bases, say A and T, is chemically weaker than the bonding of A to its upright or of T to the other upright.

Splitting the ladder from one end in this way leaves 'naked' bases waiting to be paired up. Another enzyme, DNA polymerase, assembles the requisite complementary bases, together with bits of a new upright, alongside *both* of the separated strands. So on completion of the process we have two ladders both identical to the one with which we started. What DNA polymerase does in DNA replication is clearly very similar to what RNA polymerase does in protein synthesis. Both make a nucleic acid strand complementary to a pre-existing one. In one case, the complementary strand is a DNA one, and it remains stuck to the 'template' strand. In the other case, the complementary strand is an RNA one which migrates into the cytoplasm to complete the task of making a protein. The main differences between DNA and RNA are (a) that the sugar in the 'uprights' is ribose in RNA and deoxyribose in DNA, as noted in Chapter 2, (b) that RNA molecules are usually single-stranded, in contrast to the double strand of the DNA 'ladder', and (c) that in RNA a base called U (for uracil) replaces T.

Basic genetics

Originally, geneticists were concerned solely with the transmission of characteristics of parents to their offspring. This early genetics, now often referred to as 'transmission genetics', was located in between the individual and population levels of life (see Figure 1), since its focus was essentially on the *family*. Nowadays, however, genetics has expanded to deal with everything from gene action to evolution. It thus extends through many levels of life from molecular to population. Indeed, it even extends up as far as the community, because there is a recognized phenomenon called *coevolution* wherein organisms that interact ecologically (like flowering plants and their insect pollinators) evolve jointly in response to their interaction. This wide range of application of genetics, through many levels of life, partly accounts for its dominant position within modern biology. I shall have a little more to say on this matter later.

Because genetic topics are scattered through most of the levels of organization that we have already examined, we have actually done plenty of genetics already! For example, we have looked at kinds of reproduction, chromosome behaviour in mitosis and molecular aspects of gene action. The purpose of this section is basically to connect up these various genetic bits and pieces. I also have to introduce some basic genetic terminology that appears in later chapters.

In order to do this, I want to focus our attention on the nucleus of a single cell of the fruit-fly *Drosophila melanogaster*, which is much used by geneticists. It does not matter *which* cell, as they are all the same in terms of their chromosomal complement (except for the gametes). Nor does it really matter that I have chosen *Drosophila* rather than man, or a snail, or something else. The *Drosophila* nucleus is easier to picture, since there are relatively few chromosomes, but the *principles* at stake are widely applicable among animals and plants.

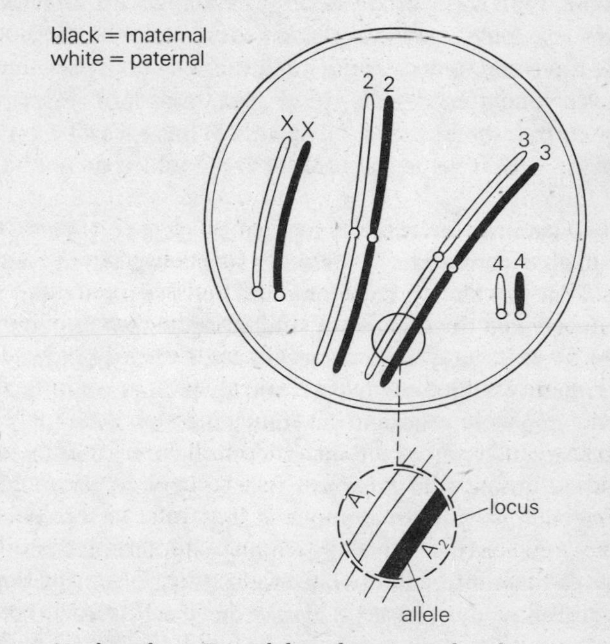

Figure 5. Nucleus from *Drosophila melanogaster*, female.

One such nucleus is shown diagrammatically in Figure 5. We shall go through this diagram bit by bit, looking at its various features. Like all higher organisms, *Drosophila* is diploid, meaning it has *two* copies of each chromosome, one derived from its father and one from its mother. *Drosophila* has only four chromosomes, each with its paternal and maternal copies, as shown. (Do not confuse two copies, at this gross level, with two helices at the molecular level. Each line in Figure 5 would be seen to contain a double helix if it were magnified sufficiently.) As in many other higher organisms, one of the pairs of chromosomes is the sex-chromosome pair. The individual shown has two X-chromosomes, which means it is female. One X and one Y is what we would see in a male. (The same is true in humans.) In *Drosophila* the sex-chromosomes are sometimes called the 'no. 1' pair, so the other chromosomes (called autosomes to distinguish them from sex-chromosomes) are given the numbers 2, 3 and 4, the last of these being much shorter than the others.

A chromosome is essentially a long DNA molecule, although it is considerably folded and twisted, and packed together with some proteins. Any short stretch of the molecule, between 'punctuation marks', is a gene. Genes, as we have seen, make proteins. One of the long chromosomes (no. 2 for instance) has several thousand genes, along with other sections of the DNA which do not have a genic function.

Let us concentrate on two different genes – one on chromosome 2 and one on chromosome 3 – which make two different enzymic proteins. Enzymes tend to have long and horrible names, but they nearly always end in -ase, so we shall imagine two hypothetical enzymes called 'onease' and 'twoase'. Even if these both have the same number of amino acids (say 100), which is unlikely since they are completely different enzymes made by different genes and catalysing different reactions (about which we don't need to know), their amino-acid *sequences* will be very different. If you run along each enzyme to amino acid 'no. 37' – or to any other position – the actual kind of amino acid found there will probably be different in the two proteins. Obviously, this implies a high degree of difference in base sequence between the two genes.

Does it follow, then, that if we look at the gene for 'onease' in different cells of our fly, or in lots of different flies of the same species, that the base sequences of those genes (and so the amino-acid sequences of the proteins) will be identical? Well, in an ideal world they might be, but in reality the sequences will not all be identical, because molecular accidents happen, and occasionally the base sequence of a gene is slightly altered. An alteration of this kind is called a *mutation*, and the simplest kind of mutation is where only one base is affected – let's say an A is changed to a G at the beginning of a particular triplet code somewhere in the middle of our gene. This will mean that, at a corresponding point in the onease enzyme, one amino acid is replaced by another. The altered protein is still close enough to the original to be considered as onease and still to catalyse the reaction that onease is supposed to catalyse, though it may perform this task slightly better or slightly worse than the original form of the enzyme. Sometimes mutations of this sort, as well as more drastic ones which result in a complete cessation of protein production by the gene concerned, are brought about by physical agents like X-rays or by exposure to noxious chemicals. But often they occur for no apparent reason ('spontaneous' mutations), just as a car will sometimes break down without any external cause such as collision with another vehicle.

As we have seen, the process of DNA replication usually involves the construction of an exact copy of the pre-existing DNA molecule. This is true even if we are copying a mutant DNA. Thus all the cells descended from a cell in which a mutation occurs will themselves carry the mutant form of the gene. If the original cell in which the mutation occurred was a somatic cell (muscle, nerve, etc.), the effect will be limited to a 'clone' of daughter-cells. But if the mutation occurs in the germ line, the mutant gene and its effects are carried on to all subsequent generations – or at least until a further mutation occurs. Germ-line mutations are thus the material of evolution; and we shall look at their spread through populations later, in the chapter on natural selection.

The site at which any particular gene occurs on a chromosome is referred to as a locus (literally, 'place'). Let us concentrate on one particular locus – locus A in Figure 5 – which is concerned with the production of one particular protein. Suppose that we start with an ideal-world situation where all copies of locus A in a population of *Drosophila melanogaster* are identical. If a germ-line mutation occurs, producing a mutant form of the gene that makes a protein which is slightly superior to the original one, that mutant gene will begin to spread through the population by natural selection. When we have two different kinds of a gene, as in this situation, we call them alleles. In this case we have two alleles of locus A, which we can call A_1 and A_2. These symbols, then, just stand for very slightly different proteins (for example, different versions of 'onease'). We don't think of an original 'true' version and an aberrant mutant one (which might be coded A_o and A_m), because evolution involves a continual, albeit slow, turnover of alleles, and there is no one ideal or true allele at any locus.

Half-way through the process of evolutionary replacement of the old allele A_1 with the new allele A_2, we have a situation where plenty of copies of both alleles are floating around in the population. Any individual can have the genetic consititution (or genotype) A_1A_1 or A_1A_2 or A_2A_2, depending on what it received from its father and mother. Of the three genotypes given, two of them are called homozygotes; the other (A_1A_2) is called a heterozygote, because the two alleles it possesses at locus A are different. What a genotype actually produces is referred to as a phenotype. For example, if the enzyme made at locus A catalyses a reaction that gives rise to an eye pigment, then A_1A_1 flies and A_2A_2 flies will have differently coloured eyes, that is, different phenotypes. Sometimes the phenotype resulting from the heterozygote (A_1A_2) is intermediate between the other two; sometimes it is the same as one of them – say A_1A_1. If this is the case, we say that A_1 is the dominant allele and A_2 the recessive one.

Two final terminological points will suffice here. First, 'genotype' refers to an individual's genetic constitution at one locus only. If you want to refer to an organism's *overall* genetic constitution, the word to use is 'genome'. Second, and rather ironically, it may now be clear that 'gene' is ambiguous. Sometimes it means locus, sometimes allele. You thus have to take care that, on meeting this apparently most basic term of genetics, you don't misinterpret it. In the context of the 'gene-frequency' of population genetics, gene means allele.

The only thing that we now have to deal with before leaving genetics is the all-important and unique type of cell division that occurs in the production of the germ-cells – meiosis. Meiosis has in fact two stages occurring in series (one after the other). The second of these is very much like mitosis, so we won't bother dealing with it here. Rather, we shall concentrate on the first stage of meiosis, which is where the crucial halving of the genetic material occurs – the halving that balances the later doubling when sperm and egg cells fuse in fertilization.

In the context of the *Drosophila* cell illustrated earlier, what happens in meiosis is that the two 'homologous' chromosomes of each pair – that is the paternal and maternal members of the pair

– separate and go into different daughter-cells, as shown diagrammatically in Figure 6. It is absolutely vital that we note that this is *not* what happens in mitotic divisions. There, each member of a pair behaves *independently*, and never pairs up with its paternal or maternal counterpart. What gets dragged apart into mitotic daughter-cells are the duplicated copies of every *one* chromosome. Because, in meiosis, it is the homologous chromosomes that get dragged apart, we find that at the end of the crucial first stage of meiosis, we have *half* of the original number of chromosomes (one from each homologous pair) instead of the same number. What we have done is to produce a 'haploid' germ-cell from a 'diploid' germ-cell precursor. When two haploid germ-cells fuse, in fertilization, we then re-create a diploid cell, namely the zygote. The only complication to this basic picture that is worthy of mention is that paternal and maternal homologues sometimes exchange bits of chromosome before going their separate ways. This process – recombination – is important because it, like mutation, helps to generate evolutionary novelty.

Since genes are carried on chromosomes, their behaviour, as we go from one generation to the next, is dependent to a large extent on what the chromosomes do. Yet despite this, the early chromosome workers did not come up with a general scheme of inheritance. It was left to Mendel, who was not a student of chromosomes, to do so. That he produced his abstract scheme without reference to the chromosomes makes it all the more remarkable, and we shall examine Mendel's penetrating insight later.

Kinds of organism

So far in this chapter, I have adhered to the horizontal way of splitting the living world, and have largely neglected the vertical. That is, I have tended to treat all organisms as equal and to talk about everything from the growth of their populations to the molecular structure of their genes as if it did not matter whether we were dealing with man, mouse, mollusc or microbe. This is an appropriate enough approach in a book which seeks to expose the unifying principles of life rather than its superficial diversity. Nevertheless, the book would be lacking if I did not at least briefly

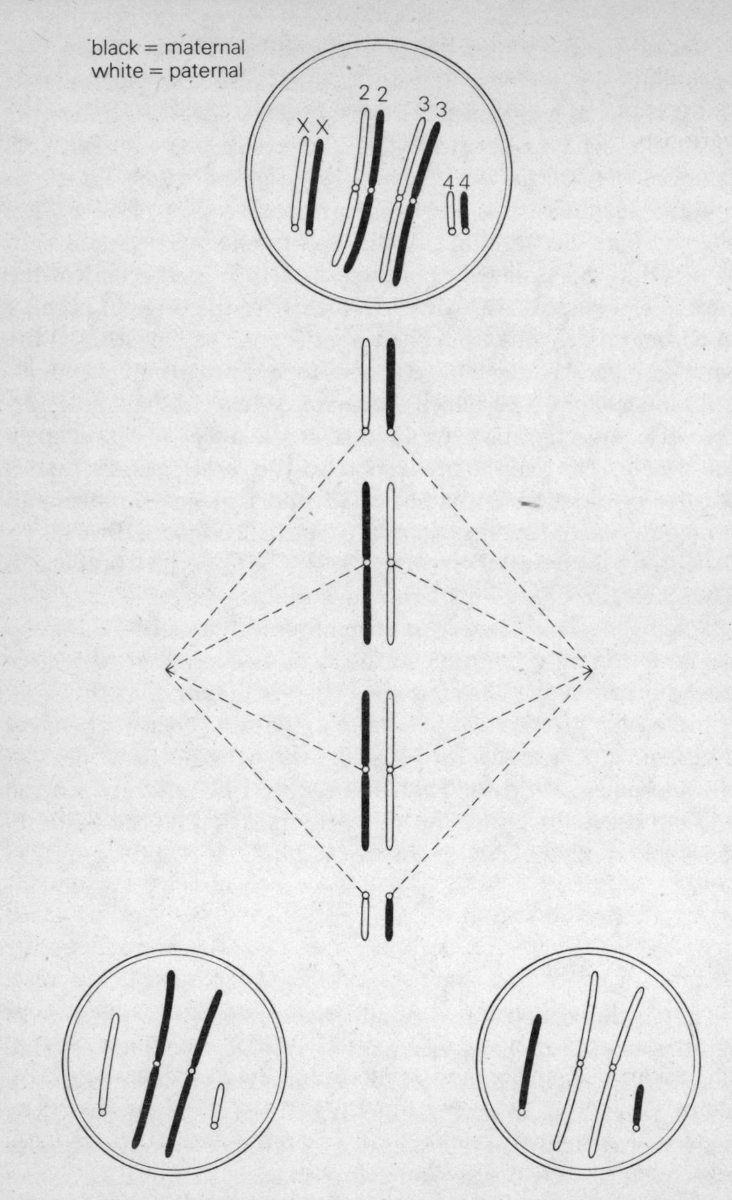

Figure 6. A simplified view of meiosis. The allocation of paternal and maternal copies to daughter-cells is random, so that each daughter-cell ends up with some of each.

allude to the diversity of *kinds* of organism that exist, partly because it is a necessary part of 'basic biology'. There is another reason also, though: a study of this diversity itself (which can be variously referred to as taxonomy, classification or systematics*) reveals that diversity is not an entirely superficial phenomenon and that it is underlain by a kind of order which is separate from the other kinds of order I shall discuss later – though there is obviously a close link between taxonomic and evolutionary order.

The pattern of order revealed by taxonomic study is best described by a phrase Darwin often used – 'groups within groups'. We all recognize certain groups of related kinds of organism, for example mammals and flowering plants. In both of these examples we can see larger groups to which the named ones belong (vertebrates, plants), and smaller ones which are subsets of them (man, daisy). Biologists believe that this sort of pattern has arisen because living organisms have been produced by evolutionary divergence from a common ancestor; and indeed, as we shall see, there can now be little doubt about this. However, it is worth pointing out for the philosophically minded reader that a pattern of groups-within-groups characterizing a set of existing objects does not *necessarily* imply an evolutionary production of those objects. Buildings or motor vehicles could easily be classified in a groups-within-groups scheme. Yet this does not mean that one sort of building, or vehicle, literally evolved into another.

The most fundamentally distinct groups that we recognize in the living world are not, as the layman may be tempted to think, plants and animals. Rather they are a group called eukaryotes, which includes both plants *and* animals, and a sister-group called prokaryotes, which consists largely of bacteria. The eukaryotes are characterized by a complex cell like that shown in Figure 3; the prokaryotes have much simpler cells which lack proper nuclei and organelles. They are also haploid rather than diploid. Within the eukaryotes, the biggest split is into kingdoms. Here we have the animal and plant kingdoms, but we need at least one additional kingdom to include those organisms such as fungi which are not really animals or plants.

Exactly what smaller groups we invoke depends to some extent

* The subtle distinctions between these three terms are discussed by G. G. Simpson in *Principles of Animal Taxonomy* (Columbia University Press, 1961).

on which kingdom we are dealing with. In the Animal Kingdom, the groups we find, going from large to small, are phylum, class, order, family, genus and species. The human species belongs to the genus *Homo*, the family Hominidae, the order Primates (which we share with monkeys and apes), the class Mammalia and the phylum Chordata. The last of these is just the vertebrates plus a few close allies.

What we have here, of course, is a hierarchical system of ordering the living world. The hierarchy starts with a bifurcation into two groups, the pro- and eu-karyotes, and ends with the 'terminal twigs' of around three million species. Connecting this with the evolutionary process, the idea is that the more distant any two living species are from each other in the hierarchy, the longer ago their ancestors diverged. The point of origin of the whole process, looking at it as an evolutionary one, is the origin of life on earth, which occurred some time between 3.5 and 4 billion years ago. Despite its importance for us, I have little to say in this book on the origin of life. This is because it was indeed a *unique* event and, while one can in a broad sense 'theorize' about a unique event in the distant past, it cannot be the basis for a general theory of the kind with which this book is concerned. A general theory about the origin of life will be possible only when we discover many more biospheres!

One advantage of a hierarchical view of the living world is that it puts paid to a common fallacy perpetuated by some non-biologists, namely that one living species has evolved from another – for instance man from chimp. Since all species change to some degree over any reasonably long period of time, no living species can ever be the ancestor of any other; rather, both diverged from a common ancestor. One of the descendant species may of course have been much more altered from the form of the common ancestor than was the other descendant, and indeed that is the case in the man/chimp example. Thus, isolating this particular small part of the evolutionary process (and using H for human, C for chimp and A for ancestor) we see that

H
↑
C is wrong;

C ↖ A ↗ H is better; and

C ↖ A ⟶ H is probably

nearest to the truth. (The vertical axis here is time and the horizontal is some measure of the difference between the organisms concerned.)

I hope the non-biologist is now at a stage where he feels that this label no longer applies to him and that he can confidently embark upon a study of unifying biological theories, always remembering that he can return to this chapter to jog his memory whenever this turns out to be necessary. And make no mistake about it – it *will* be necessary. No one without prior biological knowledge could possibly take in all of the facts presented in this chapter and still have a clear recall of all of them by, say, the middle of Chapter 12. Returning to the basics from time to time is an act of realism, not of failure, and if you remember nothing else from this chapter, please remember this!

4 A SCHEME OF HOW OUR THEORIES SHOULD FIT TOGETHER

The British biologist Sir Peter Medawar has described scientific research as 'the art of the soluble'.* The idea is basically that researchers do not bite off more than they can chew. In order to make progress in biology (or in any other science for that matter), an individual researcher must choose a question which is small enough for him to have a reasonable chance of obtaining an answer to it during his lifetime – or, in our increasingly hard-pressed economic climate, during the tenure of a particular research grant. Thus biological researchers do not set out to attempt a scientific exposition of the nature of life (even though this is what they are ultimately interested in) because it is hard to know where to begin, and the 'question', if one can call it that, is far too complex for any individual to solve. So the researcher focuses instead on a particular process of life, which he perhaps finds more interesting than others, in the hope of making some headway.

Even the concentration on a particular process – inheritance, for example – still leaves the researcher with far too great a range of questions. In order to do experiments from which he may (or may not) acquire some meaningful answers, he must become yet more restricted in the breadth of the questions he asks. If Mendel had never asked any more specific questions than 'how are characteristics of organisms inherited?' we would probably never have heard of him. That his name is a household word is due to his asking the restricted but experimentally *answerable* question: how are certain characters, such as height, inherited in garden pea plants?

* A book by Medawar with this title was published by Methuen in 1967.

It is because researchers are forced to ask very specific questions such as this that biological research often appears trivial to the layman. The fact that some biologists spend most of their working lives dealing with the height of pea plants or the number of bands on snail shells is a source of incredulity for someone unaware of the way that the body of biological knowledge has been (and is being) built up. It is easy to sneer at the 'over-specialization' of the practising scientist, and this is often done by those who cannot fathom why biologists are interested in such narrow and apparently worthless topics. What such a critic fails to recognize is that most biologists pursue narrow research projects because of their wish to pursue the art of the soluble and not because of the narrowness of their interests. Indeed, the 'typical biologist', if such a creature exists, is interested in a far wider area of biology than a superficial view of his actual experimental work suggests.

The advantage of potential solubility that derives from the study of a limited system is accompanied by a potential pitfall: it may be that the particular system chosen for experimentation has special features which make it unrepresentative of any broader class of system. If this is the case, any answers that the experimenter finds cannot be generalized. Unfortunately, whether or not this is so is often apparent only in retrospect, and thus the choice of a particular species and particular topic for investigation carries with it a risk that the information acquired will be unimportant. But then again, if such a choice is not made, the experiments cannot even begin. As we have seen, Mendel would never have achieved his (posthumous) fame had he not asked specific questions. But had he chosen to ask not about inheritance of size in peas but about inheritance of the direction of coiling in snails, he would probably have failed to come up with a theory of inheritance, and if he had, and had proposed it as a general theory, he would have been wrong. (The reason for this will become apparent later.) What all this tells us is that good science requires the boldness to choose a system for study based on limited information, the sense to use what information is available to choose wisely, and a good deal of luck to boot.

What the bold, wise and lucky biologist, such as Mendel, is able to achieve is not just an answer to his specific question, but the answer to a much broader one which was probably what he was interested in in the first place. The sequence he has followed is not broad question → broad answer, which unfortunately is not usually possible, but rather the more complex sequence: broad question → narrow question → narrow answer → broad answer. The final step in this sequence is the act of generalization. However, even for a highly successful biologist such as Mendel or Darwin, the generalization has to stop somewhere. No experimental result has yet been generalized into a complete scientific exposition of the nature of life, and it is most unlikely that any ever will be. Thus what we end up with, at best, are theories covering major life processes such as development, inheritance and evolution.

Given, then, that we have (or in some cases will have) separate theories covering the distinct major processes of life, do these fit together into some sort of meaningful complex? In other words, although we do not have a single, unified 'theory of life', do the relatively small number of general theories about life processes that we do have provide, in some sense, a unified and self-consistent overall picture?

In order to answer this question we need to do two things. First, we need to identify where particularly strong links between particular pairs of theories should exist. Then we need to examine these linkages to see whether they reveal a clash or a compatibility between the existing theories in related areas of biology. An outline of the linkage pattern among areas of theory is given in Figure 7. Whether or not existing theories are mutually compatible cannot be examined in detail until after the theories themselves have been presented individually (Chapters 5–12); so this will be postponed until Chapter 13.

Genetics: the core of biology?

Genetics is placed in a central position in Figure 7 for two reasons. First, because it has more links radiating from it than any of the other areas. Second, because the present trend in the biological sciences is for genetics to infiltrate into areas which were previously considered to be outside its remit. Indeed, geneticists now address such a range of biological problems that the equating of

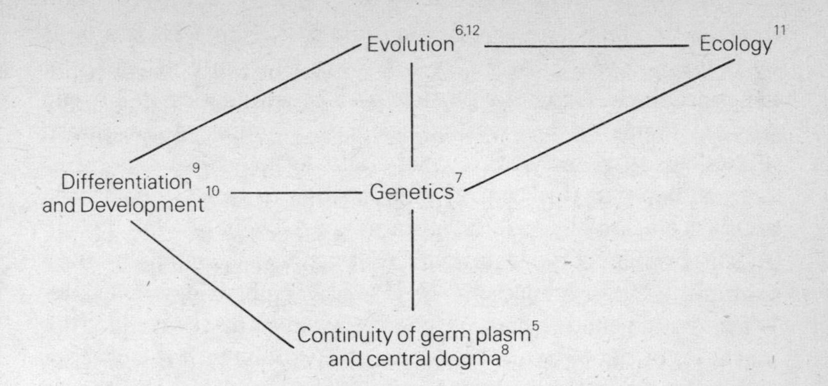

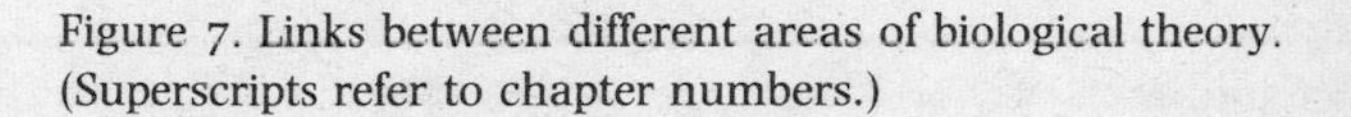
Figure 7. Links between different areas of biological theory. (Superscripts refer to chapter numbers.)

the discipline of genetics with the process of inheritance which was what geneticists first set out to study is no longer tenable. I remarked earlier that a useful label for the study of inheritance is transmission genetics (because it is that branch of genetics still dealing with the original problem – transmission of traits from parents to offspring). Other branches of genetics include developmental genetics, which has revolutionized embryology, population genetics, which is now one of the main branches of evolutionary biology, and molecular genetics, the core of the wider field of molecular biology, which is concerned with such things as gene function and control.

The infiltration of genetics into other areas of biology has not been reciprocated. Embryologists and ecologists, for example, have made little contribution to the study of genetic problems. Does this mean that genetics is in some sense the core of biology, and that genetic theories will ultimately supersede all others?

The current mood in biology, with genetics in general and molecular genetics in particular being in many ways the dominant field, might lead one to suspect that the answer is an unqualified 'yes'. In fact, 'yes and no' is a more accurate, if initially obscure, answer.

It is of course true that most if not all attributes of organisms are ultimately produced, directly or indirectly, by the genes. This includes the genes themselves, since any particular set is produced by replication of a pre-existing set. So in this sense at least, the genes are the fundamental particles of life – analogous, in a broad way, to atoms in the physical sciences. (Neither of course are indivisible, as atoms were once thought to be!)

In addition to this central position of the genes themselves, another clue to the present dominance of genetics lies in the informativeness of the techniques that geneticists employ. A good example is the comparison of classical embryology with developmental genetics, both of which are aimed at unravelling the mysteries of egg-to-adult development. By looking at the effect of certain mutations (defective genes) on development, we can begin to understand what purpose the normal version of the gene serves. Such a technique clearly surpasses the embryologist's scalpel in its ability to yield information on the causative steps involved in the developmental process.

The central position of the gene and the utility of genetic techniques have, however, caused some geneticists to adopt an unacceptably narrow view of biology. One area in which this is apparent is the field of molecular evolution. Many geneticists interested in evolution have turned their attention inward and focused on the evolution of the genetic material itself – the evolutionary increases and decreases in the amount of DNA, the evolution of strange mobile genes which 'jump' from place to place unlike the more typical static Mendelian gene, and other such molecular evolutionary processes. What such geneticists seem to forget is that the *important* aspects of evolution are those that relate to the structure, function and behaviour of whole organisms. Some of the molecular evolutionary processes currently under study bear little if any relation to these things and hence to *adaptive* evolution. This is not to say that evolution of the genome itself is of no consequence, but rather that it must be seen in perspective against, and in a way as subservient to, a much broader evolutionary backcloth.

It may now be apparent that the answer to our earlier question is indeed literally 'yes and no'. Yes, genetics is in some ways at the core of biology; but no, genetic theories will not ultimately supersede all others. Theories that started out without a genetic component may find that they come to develop one. But this is quite different from their being *replaced*. I hope this view, which acknowledges the potential of newer molecular and genetic approaches without neglecting the advances made by older, more descriptive ones, is agreeable to a majority of biologists. Certainly it seems preferable to a stick-in-the-mud refusal to accept new approaches or an arrogant dismissal of the considerable knowledge acquired with older ones.

On not neglecting the wood

After this chapter the unsuspecting reader will be plunged into a series of narrower, more detailed accounts of particular biological theories. Whether a biologist or not, he may feel surrounded by the tall, informational trees of the theory of Weismann, or Mendel, or whoever. Because of the importance of retaining an overall view of things, this more general part of the book ends with a plea to keep returning to the wood. If you feel that, in the middle of Chapter 9 for example, you have lost an overall perspective, re-read the short introductory chapter, which expounds the central theme of order, or re-inspect Figure 7, which paints an overall picture of theoretical biology. Quite apart from the personal satisfaction of maintaining a broad view of biological theory, such a view is helpful in rebutting the attacks on rational thought which periodically emerge (particularly in the United States) from the anti-biological 'creationist' lobby. The scrambled ideas emerging from this quarter may seem like very ineffective missiles; but the blinkered biologist pursuing his own speciality to the exclusion of all else may one day look up to find the citadel he took for granted lying in ruins because no one had bothered to defend it.

5 THE CONTINUITY OF THE GERM PLASM: WEISMANN'S THEORY

One of the most unfortunate of the great biologists was the Frenchman Jean Baptiste Lamarck. Although he was one of the few pre-Darwinians to propose that the present array of species was brought into existence by evolution rather than by independent creation, Lamarck is rarely remembered for this. Instead, his name has become inextricably linked with the 'inheritance of acquired characters', a mechanism of heredity which we now know does not occur. An example which is often given to explain this mechanism, and to distinguish it from natural selection, is the giraffe's lengthy neck. If a primitive giraffe kept striving to reach food on high branches, its neck might gradually stretch a little. If such an acquired character is inherited, the offspring of our hypothetical giraffe will be born with a slightly longer neck. This process could be compounded over many generations giving, ultimately, the long neck of today's giraffes from a much shorter-necked ancestor.

Darwin's advance over Lamarck was, of course, that he postulated natural selection (to be considered in the next chapter) as the principal cause of evolution. Unlike the inheritance of acquired characters, natural selection has now been confirmed by a vast array of experiments in which it has been well documented and quantified. Ironically, though, and in contrast to the recent emphasis of this difference between the two evolutionists, Darwin did not completely reject the inheritance of acquired characteristics – rather, he relegated it to a subsidiary evolutionary role. It was left to the German naturalist August Weismann to take an axe to this idea and, more generally, to the idea that changes occurring in the body of an organism were in some way incorporated into information in its germ-cells and thus passed on to its progeny. Weismann's insistence that these ideas were false gained acceptance towards the end of the nineteenth century. More recently, molecular biologists have re-expressed Weismann's main proposition as their 'central dogma', which will be examined later.

Organisms as lines through time

Living organisms, of whatever sort, are exceedingly complicated things, and to make life easier for themselves, biologists represent them in various abstract ways. The aim of such abstractions, as with any in science, is to dispose of information which is irrelevant for the task at hand and which is therefore liable to be distracting rather than helpful. For example, field ecologists interested in patterns of distribution in space often represent organisms as dots. This is, arguably, the highest degree of abstraction that can be achieved. Nevertheless, it is perfectly sufficient if all that is of interest is whether a population has some definite distribution in space (it might, for instance, occur in 'clumps') or whether the organisms are simply distributed at random. Indeed, for an ecologist interested in the spatial distribution of a population of large mammals, such as wildebeest, inhabiting an open plain, the quickest way to get an accurate picture of the distribution at a particular moment in time is to go up in a helicopter and take an aerial photograph. If taken from a substantial height, this will essentially show the organisms of interest as dark dots on a paler background. In a situation like this, the loss of detailed information on wildebeest appearance, such as would be apparent from a close-up photo of a single animal, is not just harmless but is actually helpful – with the proviso that there is not a second species of large mammal inhabiting the same plain which would also show up as dark dots!

This is just one of many abstractions which biologists use. The essential point about it is that it retains only the information about organisms that is currently needed, which in that example was their position in two dimensions. The abstraction required to consider Weismann's theory and alternatives to it is different; here, we are interested in patterns of temporal continuity rather than patterns of spatial distribution – so an organism ceases to be a dot and becomes a line through time.

Chickens, eggs and immortality

The simplest type of organism to represent in this way is a (hypothetical) immortal one which exists throughout time and gives rise to no offspring. This is shown on the left-hand side of Figure 8. By breaking the line up into finite pieces, we arrive at an abstraction of a series of consecutive generations of a mortal organism. If we wish to show that each generation gives rise to the next not directly but via germ-cells, an alternating dot–dash line will suffice. Now although this representation has some problems, it is the most realistic of the three so far given.

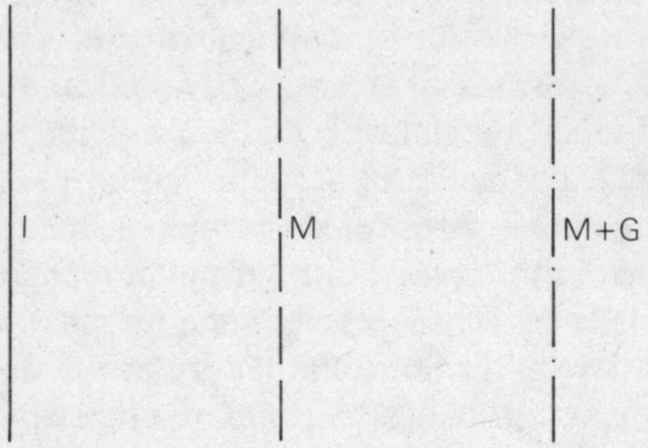

I M M+G

Figure 8. Organisms as lines through time. (I = immortal organism; M = mortal organism; M + G = mortal organism reproducing via germ-cells.)

It will be helpful for our current purposes to replace this representation with C (for chicken) and E (for egg). In that case, our dot–dash pattern takes the form shown at the left-hand side of Figure 9. It will be clear that this is a decidedly female-biased view; roosters are, after all, required at certain stages. However, the principle of omitting information that is temporarily irrelevant is again being applied. Weismann's view of the relationship between chicken and egg (or generally between body (soma) and germ-cell) can be seen, in its most extreme form, by reorganizing the pattern of C–E linkages to a branching form, as shown in Figure 9.

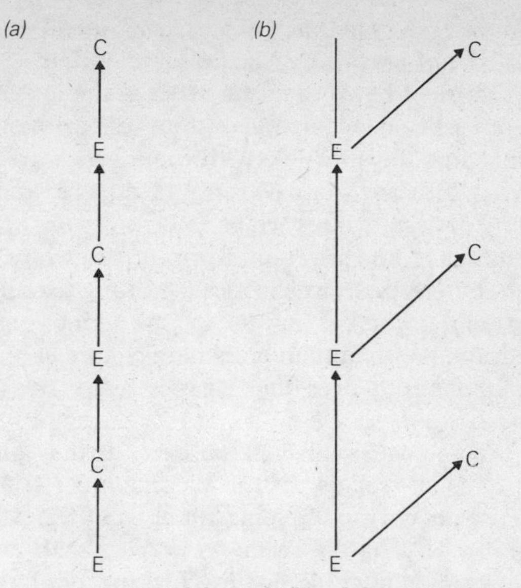

Figure 9. (*a*) A re-statement of the right-hand diagram of Figure 8. (*b*) The Weismannian view.

If we interpret the arrows as indicating flow of information, it is apparent that, in Weismann's picture of things, changes occurring in the informational content of a germ-cell affect both the body derived from it and future generations, whereas changes occurring in the body itself represent a hereditary and evolutionary cul-de-sac, and have no consequence beyond the individual in which they occur.

The bearing of the branching C–E pattern shown on Lamarckian and Darwinian evolutionary mechanisms is clear. For the inheritance of acquired characteristics, we need to have an arrow from C to E in each generation as well as an arrow from E to C. Since no arrows of the former kind exist in Weismann's scheme, the Lamarckian mechanism is precluded. Natural selection, on the other hand, is perfectly compatible with 'Weismannism'. The kind of mutation upon which Darwinian natural selection acts to produce long-term evolutionary change is the germ-line mutation. Now in the Weismannian scheme, such a mutation, appearing at any particular E-point in the diagram, will potentially affect the C derived from that E as well as all later C's and E's. This is compatible not only with natural selection but also with our growing picture of how organisms work at the molecular level.

So far, our scheme of temporal continuity is working out rather well in that it is incompatible with the inheritance of acquired characters, which we know does not take place, and it forms a background pattern against which to consider the accepted processes of mutation and selection. However, as always, we cannot merely be content that our theories are fitting together well; we must also inquire whether they are in tune with biological reality. Thus, with the Weismannian branching picture of the pattern of organismic continuity over time, we must ask: is this an accurate abstraction?

There is some cause for concern here, as may already have occurred to some readers. In the Weismannian scheme, germ-cells give rise not only to the soma but also to the next generation of germ-cells. This hardly seems in keeping with everyday experience. Eggs are, after all, laid by chickens, not by other eggs.

This observation would, at first sight, seem to make the Weismannian scheme untenable. However, if we delve a little more deeply into the relationship between soma and germ line in a developing organism, we shall see that Weismann and reality are not so far apart. One organism whose development has been studied extensively is the geneticist's favourite fly – *Drosophila*. The life-cycle of this fly is fairly typical of higher animals in that, following sexual reproduction, the fertilized egg develops, ultimately, into the adult organism. Of course, since *Drosophila* is an insect, the developmental process is a multi-stage one, with larva and pupa intervening between egg and adult, but this complexity need not concern us here.

As the fertilized egg multiplies, different groups of cells, which will eventually give rise to different tissues, become recognizable. In *Drosophila*, the first such group is a cluster of 'pole cells' which appear as a distinct entity a mere two or three hours after fertilization. (The complete developmental process takes at least ten days in *Drosophila*.) Once formed, the pole cells remain distinct from other developing cells. Ultimately, they give rise to the adult fly's germ-cells. What this means is that, although there is a very short phase in early development wherein no distinction can be made between soma and germ line, this phase is rapidly replaced with a clear separation, which then persists throughout the subsequent life of the individual. So Weismann's view of the living world is not so inaccurate after all, and, in a sense, eggs very nearly do lay eggs. The Weismannian branching scheme given in Figure 9 could, of course, easily be modified to take account of the short phase in early development when no germ/soma distinction is apparent. (If you want to devise such a 'modified Weismannism', it will help to use X, or some other symbol, to represent the 'joint' stage of development.)

The pitfalls of generalization

While the biologist's ultimate aim is to produce theories of the living world that are as general as possible, the diversity of organisms (our starting point in this book) presents considerable obstacles to the realization of this aim. Most biological theories are not universal – the best we can usually hope for is a theory that holds in the vast majority of cases. In many of the following chapters in which individual theories are described, exceptions are known. For example, not all characters of all organisms display standard Mendelian inheritance, and not all cells differentiate entirely through the switching on or off of genes. It could be argued that Darwin occupies a special place among biological theorists in that all populations of all species are subject to natural selection. However, even here, the universality disappears if we phrase our generalization differently. The statement 'all evolution is a result of natural selection' is certainly wrong, as we shall see.

The problem of attempting to generalize in the face of biological diversity affects Weismann's theory just as it does the others. In this case, the problem arises because of the different developmental patterns, life-cycles and modes of reproduction displayed by different species. As we have seen, the details of *Drosophila* development largely support Weismann's view. But what of other species? In fact, *Drosophila* is reasonably representative in this respect of 'higher animals' in general. (This phrase, incidentally, is normally used not just for vertebrates, as some people mistakenly think, but also for many invertebrates, including insects, molluscs and worms. The contrast is with primitive animal taxa such as protozoans, jellyfish and sponges.)

The organisms to which Weismann's scheme does not apply so well are some plants and lower animals, together with many of the life-forms that do not fit into either the animal or plant kingdoms, such as bacteria. The most severe departure from Weismannism is presented by unicellular organisms (of whatever group) that reproduce by the asexual process of binary fission, in which the cell simply splits into two more-or-less identical daughter-cells. Since it is impossible to say that one daughter-cell is 'older' or represents the remains of the parental cell, what we have here is something close to potential immortality, at least of the genome containing the information from which each replicate cell is built. The qualification of 'potential' is necessary because in reality any such cell line will eventually die out. The biosphere, after all, will not last for ever.

There are just two final points that should be made in relation to the departures of some types of organism from a Weismannian pattern of temporal continuity. First, in no case is there a well-established arrow from C to E, of the sort that would be necessary for the inheritance of an acquired character. Some recent experiments performed by the Australian immunologist E. J. Steele have suggested the presence of just such an arrow, but repeats of these experiments by other workers have failed to confirm Steele's findings. Second, I do not mean to imply that Weismann himself was unaware of the variation in life-cycle found in the living world and the fact that it cannot be constrained to fit a single, simple pattern. Indeed, Weismann drew attention to the potential immortality embodied in the binary fission reproductive system discussed earlier. The way science works, earlier distinguished authors associated with certain theories or processes are often construed as pursuing their pet theory in a single-minded, even pig-headed, manner. In fact, while some such scientists can always be found, most people whose ideas developed into general biological theories were well aware both of variations which did not neatly fit their theory, and problems of the theory itself. This was certainly true of Weismann, and also of Charles Darwin, whose theory of natural selection is discussed in the following chapter.

6 EVOLUTIONARY THEORY: DARWINIAN NATURAL SELECTION AND ALTERNATIVE MECHANISMS

There is no such thing as 'the theory of evolution'. The use of this label (which is about as vague as 'the meaning of life') has served only to confuse. In particular, it has led to the confusion of the radically different issues of *whether* evolution has occurred and *how* it has occurred. I shall take the first of these issues for granted in this chapter and shall concentrate on how evolution has taken place – that is, through what mechanism it has been, and is being, driven. Certainly, the question of whether evolution has occurred at all is not at issue among biologists, the vast majority of whom (probably well over 99 per cent) are convinced of its reality. A recently formed group of taxonomists – the 'transformed cladists' – has muddied this uniform picture slightly by disposing of evolutionary relationships as a guide to classifying existing organisms. However, even this retrograde step, as it seems to the rest of the biological community, does not involve proposing that evolution does not occur. Rather, the transformed cladists are merely claiming that it is possible to devise systems of grouping living organisms which do not use evolutionary relationships to achieve the groupings, just as it is possible to devise groupings that do use these very relationships. So even those biologists who are the most dismissive of evolution are not, for the most part, denying its reality; they are just denying its utility for certain practical purposes.

Turning to how evolution takes place, the first proposed evolutionary mechanism to gain general acceptance among biologists was of course natural selection. This was described by Charles Darwin and Alfred Russel Wallace in their papers to the Linnean Society of London in 1858 and, at much greater length, in

Darwin's classic *On the Origin of Species by Means of Natural Selection*, published the following year. While natural selection is still accepted by nearly all biologists as *a* mechanism of evolution, it is no longer seen as *the* mechanism.* Whether the others are rare and unimportant, or whether they will eventually relegate natural selection to that category, remains to be seen. Opinions on this vital matter vary dramatically among evolutionary biologists today, and we shall look at some of these opinions later. First, though, it is necessary to look more closely at natural selection itself.

What is natural selection?

Natural selection is often equated with the 'survival of the fittest', and this has caused problems because of the different interpretations that can be made of the word 'fit'. Actually, in evolutionary theory, fitness has a fairly precise meaning, which does not correspond very well with its everyday meaning of 'healthy'. Evolutionary fitness encompasses two separate variables, namely average life-span and average number of offspring. These are averages in the sense that we think of the fitness of a particular genotype, which may be represented by very many individual organisms whose actual life-spans and offspring numbers vary considerably. However, despite this variation *within* any genotypic category, there may also be a significant difference *between* genotypes in their average fitnesses, and it is this that constitutes natural selection.

As well as life-span and offspring numbers taken individually, 'fitness' also incorporates the interaction between the two. To put it another way, two genotypes with identical average life-spans and identical average offspring numbers will not be equally fit if one of them gives rise to its offspring at an earlier age. This effect can be seen in human families; the rate of increase in population would be much lower in places where the average age of marriage is quite late (such as rural parts of Ireland) than in places where people tend to marry relatively early (as in industrial parts of

* Actually Darwin did not see it that way either!

England), even if the life-spans and actual offspring numbers were exactly the same – which they are probably not.

The actual measurement of fitness and its mathematical definition, which go hand in hand, are much more complex issues and are beyond the scope of a book of this kind. But it is worth noting that there are two rather different measures of fitness available to the evolutionary biologist, both of which take into account all three factors of life-span, offspring number and their interaction. The difference between the two available measures is that one of them, the 'net reproductive rate' (usually labelled R_o), measures the fitness of a particular genotype without explicit reference to the rest of the population in which that genotype finds itself. The other, which I shall call *relative fitness* and which is often labelled *w*, measures fitness in a comparative way, so that an organism is described only as more or less fit than others in the same population. Darwin's theory and its modern counterpart – often referred to as neo-Darwinism or the 'modern synthesis' – is entirely based on the idea of relative fitness. This seems to make sense, since evolution involves changes in the relative frequencies of different genetic variants in populations; but there may be circumstances where the idea of relative fitness is *not* applicable, and we shall return to these later.

Is natural selection a tautology?

It has been asserted by a great many authors that the idea of natural selection is in some way lessened in its worth because it is tautological. A tautological statement is a circular one and is in a way akin to a definition. Whether its circularity is a problem or not depends on what is sought from the statement at hand. For example, if we say that 'all blondes have fair hair' and expect this to tell us something important about blondes, we are clearly being ridiculous. But as a *definition*, the statement 'blondes are people with fair hair' is perfectly satisfactory. A similar distinction can be made in relation to natural selection. Those authors claiming it to be tautological are usually putting it in a form such as 'natural selection results from survival of the fittest'. Since natural selection *is* survival of the fittest, this statement is clearly ridiculous and, for that matter, tautological. But if we say that, 'within a population, genotypes differ in fitness' or, to put it another way, 'natural selection occurs', then there is nothing at all wrong; we are simply making a testable hypothesis.

Is natural selection refutable?

At the other end of the spectrum from authors claiming that natural selection explains nothing because it is merely a tautology are those who claim that it explains too much. The idea being put forward here is that any biological observation can be 'explained away' as a result of natural selection and it is difficult to prove that this is not indeed the correct explanation. Those who put forward this view are essentially saying that the idea that the 'adaptations' we see in living organisms (beaks, fins, colour patterns and so on) are due to natural selection is not refutable – and so we are not dealing with a scientific theory. This view has been criticized by the British evolutionary biologist John Maynard Smith, who has pointed out that if a series of pigment-spots on the tail of one species of fish took the form of one heavenly constellation and the patterns of other similar species were of other constellations, he would regard this set of observations as refuting the hypothesis that natural selection was the cause of the pigment-patterns. As Maynard Smith himself points out, the ridiculousness of the example only goes to show that most of our observations of real species are compatible with natural selection – that is, they do not refute natural selection as the agent of evolution. Being refutable and being refuted are quite different things. A correct scientific theory is refutable but not actually refuted, and in many respects natural selection falls into that category.

The bases of natural selection

Having disposed of the philosophical problems that are sometimes raised in relation to natural selection, we shall now turn to look at those biological phenomena which are necessary if natural selection is to occur. In fact, it is possible to describe a set of phenomena which are, in the language of the mathematician, 'necessary and sufficient conditions' for natural selection. In other words, natural selection will not occur *without* these phenomena, but if they are present in a system then natural selection *must* occur – that is, it becomes a logical necessity.

The list is very short, and consists only of reproduction, variation and inheritance. The first two are not in the least problematical – we all know very well that living organisms reproduce; equally, all known types of organism exhibit variation from one individual to the next, our own species providing a particularly clear example. Inheritance was a problem for Darwin and his contemporaries, for while family resemblances suggested that variation was indeed inherited, the case was weakened by the lack of a known mechanism. This problem is now completely solved, and indeed the basic mechanism of inheritance was worked out in the last century by the monk Gregor Mendel. Ironically, Mendel's work was going on at around the same time as Darwin's (Mendel's paper was published in 1866, only seven years after publication of *The Origin*) but Darwin was unaware of it. The significance of Mendel's work eluded the scientific community until its 'rediscovery' in 1900, but his ideas have since then turned into one of the central theories of biology and will be discussed in the next chapter.

Since we now know not only that all living organisms reproduce and vary, but also that they exhibit inheritance (operating under a very general set of rules), does this mean that natural selection must therefore occur in every living system? Actually it does not, despite what I said earlier – that reproduction, variation and inheritance were necessary and sufficient conditions. No other additional process need be considered, but we *do* need to specify an interaction between variation and reproduction. That is, some variants must produce more prolifically than others. Given this interaction, natural selection logically follows – indeed we are in danger of getting back to tautologies again here, because the state of affairs just described – an inheritable tendency for some variants to reproduce more than others – *is* natural selection!

A classic example of natural selection

At the risk of boring the reader, I shall describe in outline what has become one of the classic cases of 'natural selection in action' – that is, adaptive evolutionary change actually caught happening. (Most of our evidence for such change is indirect and takes the form of data which suggest that adaptive evolutionary changes have occurred in the past.) The example, as many will have guessed by now, is the peppered moth *Biston betularia* and in particular the phenomenon of industrial melanism whereby populations in industrial areas come to have high frequencies of melanic (darkly pigmented) individuals. The story of *Biston* provides a good starting point from which to move on to more complex issues.

What has happened over the last century or so in British populations of this moth is as follows. Before the industrial revolution had widespread effects, tree-trunks throughout the country had pale, lichen-covered, grey-green barks. The moths rested on these during daytime with their wings spread flat against the trunk. The wings were a mottled greyish colour, and thus the moths were very difficult to see and consequently suffered a relatively low level of predation by birds. As the effects of industrial pollution spread, involving the death of many species of lichen and their replacement, on tree-trunks, by a thin covering of soot, the moth became more and more conspicuous and (so we assume) suffered increasingly from predation. In the new environment prevailing in the industrial areas, the moth's coloration was no longer adaptive, since to match the background and remain relatively invisible it was now necessary to be almost black.

Around 1850, a few black moths were found in the Manchester area. Over the next century the frequency of these melanics in urban populations of *Biston betularia* increased from a fraction of a per cent to a very high level – more than 90 per cent in many populations, and reaching almost 100 per cent in a few. Thus the 'typical' form of the moth in these populations was now melanic, and the original form a relative rarity. Through this evolutionary event, the population in a sense 'caught up on' its changing environment and regained the level of camouflage it originally had.

The last three paragraphs describe the story as it is usually told to the layman. There are, however, certain additional elements that we need to examine, which are less well known.

First, the genetic basis of the melanic phenotype is known, but its full biochemical basis is not. Melanic phenotypes are caused by mutation of a single genetic locus; and the melanic mutation is dominant, so both heterozygotes and mutant homozygotes appear almost equally darkly pigmented. However, the gene concerned does not make the pigment. Presumably it makes an enzyme which is involved in pigment production, but the full details of this process are not known. Pigmentation is of course genetically fixed. That is, given a certain genotype at the locus concerned, a moth will inevitably be pale or melanic. The result does not depend on any direct environmental influence, from what the moth eats to where it perceives itself to be. There is a world of difference between genetically fixed pigmentation and the colour-changing flexibility of the chameleon.

Second, there is nothing 'magical' about the appearance of the melanic forms in the place and at the time that they were needed (industrial Manchester in the last century). One of the main contentions of Darwinism, and its main difference from Lamarckism, is that the appearance of a new adaptive variant is in no way linked to the appearance of an environment in which it will be useful. In the context of *Biston*, we assume that melanics were appearing by mutation at a rate of perhaps one in a million moths long before the industrial revolution. The mutation itself, like all others, is a sort of molecular accident. Prior to the industrial revolution, these mutant moths were acted *against* by natural selection because they were *more* conspicuous than their pale equivalents. Under these circumstances, the frequency of the melanic will be maintained at a very low level, or will oscillate between zero and a low level, as a result of the balance between mutation and selection. What happened in the last century was that this mutation turned up for the umpteenth time and happened to find itself in a now-favourable environment. Mutation and selection then worked in the same direction, and this is precisely the situation needed for evolutionary change to take place.

Third, it has been possible to demonstrate that natural selection by avian predators is indeed the cause of this evolutionary change in *Biston*. Experiments in which groups of melanic and non-melanic moths were released into urban and rural environments showed that a disproportionately high fraction of non-melanics were eaten in the urban environment, and that the situation was reversed in the rural one.

Fourth, the ultimate fate of the urban populations depends on the environment. If conditions remain the same, the end-result of the process of directional selection should be complete eradication of the less fit variant. This event is referred to as a *fixation* – sometimes, the new variant is said to have been 'fixed' in the population. However, since pollution levels have recently dropped, and the nature of the pollutants has changed, with soot being much less in evidence than before, populations in some industrial areas have begun to 'backtrack' – that is, the frequency of melanics is decreasing again. Even a population that had reached fixation may backtrack, since the pale form may recur owing to mutation or to migration from a neighbouring population. The latter alternative serves to stress that the evolutionary fate of a population is linked to what is happening in other populations of the same species.

Finally, it must be very strongly emphasized that the sort of evolutionary change observed in *Biston*, even if it ends in fixation of a new variant in a group of populations of a species, does not constitute *speciation*, that is, the origin of a new species. Almost always, individuals of the different phenotypes can interbreed quite freely, and since this is the way species are defined, populations fixed for alternative forms are most certainly not species in their own right. The use of latinized names for particular variants, such as *carbonaria* for the commonest melanic phenotype in *Biston*, is confusing in this respect. Novices in evolutionary biology are apt to think of populations of the melanic as being *Biston carbonaria*, which suggests a separate species. The correct usage is *Biston betularia* variety *carbonaria*; but it is better to avoid using formal names completely for variants such as melanics, and the general trend is for biologists to drop this confusing practice altogether. So, to conclude: natural selection does not always cause fixations, and an individual fixation will only very rarely cause a speciation.

The concept of polymorphism

Despite being told many times, industrial melanism in *Biston* still makes a good evolutionary story – especially as it is a continuing one, and new twists in the plot, like the recent drop in pollution, keep cropping up. However, a particular evolutionary story does not constitute a general evolutionary theory, and we must now begin to broaden our horizons in order to make this transition.

The first step in this broadening process is to acknowledge that what is happening in *Biston*, namely industrial melanism, is happening in many other species of moths, in other groups as well, such as ladybirds, and in Europe, Asia and America as well as in Britain. It is therefore not as restricted a story as we are led to believe by repeated description of the *Biston* case-study and omission of the others. (The reason for this is simply that we know less about the others.)

Continuing to widen our horizons, the next step is to see industrial melanism involving pale and dark forms of a species as a special case of evolution of pigmentation patterns. Many cases of evolutionary change in coloration are entirely unconnected with human modification of the environment, and involve colours and patterns that are not just light and dark. An example is the land-snail *Cepaea nemoralis*, which is very variable for both the colour and banding-pattern of its shell, with populations in different kinds of habitat often having characteristically different frequencies of the different variants. Many thousands of species of living organism exhibit variation in colour and/or pattern, land-snails and butterflies being particularly well represented.

Having seen that one case of industrial melanism is only a particular example of a more general process, and that industrial melanism as a whole is only one component of the evolution of pigmentation, we now need to make the final step (for the moment anyway) of our broadening process. We need to see evolution of pigments as a special case of evolution of any characters which vary in a discrete manner. By discrete, I simply mean a fixed number of alternative possibilities, such as pale and dark, or yellow, pink and brown. The opposite situation is where we have a continuous variable such as height or weight, where endless, negligibly different character-states grade into one another. (We shall return to this second state of affairs later on; one final point here – in some cases, such as skin colour in man, pigmentation falls into the continuous rather than the discrete category.)

In order to talk about the evolution of a diverse collection of characters whose only common theme is that of discrete variation, we need to have a general terminological framework. We refer to populations containing an appreciable frequency of two or more alternative variants as polymorphic; or, to put it a slightly different way, they exhibit a polymorphism. A population of *Cepaea* snails with 72 per cent unbanded shells and 28 per cent banded is therefore polymorphic, while one with 100 per cent unbanded is not. Exactly where the distinction is drawn, and what is meant by 'appreciable frequency', have been the subject of some debate, but usually we consider 2-variant populations where the rarer variant comprises at least 1 per cent of the population as polymorphic. Quite apart from the theoretical issues concerned, this has the practical advantage that we 'write off' as monomorphic those populations where the rarer variant is at such a low frequency that it would not show up in small samples taken from a natural population for the purpose of studying the variation.

A population that is polymorphic with respect to a particular locus can have its genetic structure described in three ways. In order to examine these, it will be helpful to think in terms of a locus called A with alleles A_1 and A_2 and individuals thus having genotype A_1A_1, A_1A_2 or A_2A_2. We shall assume that there is complete dominance, so that A_1A_2 individuals are phenotypically identical to A_1A_1's; and that we are dealing with flower-colour in a plant species, with A_2A_2 giving a white flower, the others a red one.

Using this system, we can look at the three measures of a population's genetic structure – the gene (or allele) frequency, the genotype frequency and the phenotype frequency. In each case, it is the *relative* frequency, on a scale from 0 to 1 (or 0 per cent to

100 per cent if you prefer) that is implied. Let us now consider the following small population where the number of individual plants (N) is 10; each genotype below represents a particular plant:

A_1A_1	A_1A_2	A_2A_2
A_1A_1	A_1A_2	A_2A_2
	A_1A_2	A_2A_2
		A_2A_2
		A_2A_2

Suppose we are interested in red flowers or the gene or genotypes that act to produce them. The phenotype frequency is then simply the frequency of reds (0.5); the frequency of red homozygotes (one measure of the genotype frequency) is 0.2; and the frequency of the 'red allele' is 0.35. Note that in the last case we detach ourselves from individual organisms, and count actual *genes*, dividing by the total number of genes which, since our plant is diploid, is 2N rather than N.

Having distinguished between these three variables, I would like to make two additional points. First, as should be clear from the above, the numerical values taken by the variable will be different; however, if a trend towards one end of the scale occurs in one variable – such as a trend towards increased frequency of red phenotypes because of natural selection – then the other frequencies will move in the same *direction*. Second, which variable is concentrated upon depends on the situation. If we are indeed looking at red and white flowers, or any other situation where there is complete dominance, we cannot tell heterozygotes from dominant homozygotes, and so we cannot directly measure the genotype or gene frequencies, so it makes sense to deal with phenotypes. There are other situations where one can observe all the different genotypes (such as in enzyme polymorphism, to be described shortly) and consequently one can estimate both genotype and allele frequencies. In this situation biologists tend to concentrate on the frequencies of the genes themselves. The reason for this is that genes are in a sense the 'atoms' of evolution; they are not split or fused by the process of mating, as the population goes from one generation to the next. This contrasts with genotypes where, for example, in a mating between A_1A_1 and

A_2A_2, none of the offspring is of either parental genotype – all are heterozygotes (A_1A_2).

Types of polymorphism

Early work by population geneticists on evolutionary changes in polymorphic characteristics concentrated upon rather specific sorts of genes, and their products, which were not particularly representative of the genome as a whole. Much of the work was on genes involved in pigment production, as we have already seen. Another important group of investigations was based on polymorphism of blood proteins in vertebrates. For example, the ABO blood group system in man is a genetic polymorphism, as is the widespread variation in the structure of haemoglobin, this being at its most pronounced in the S-type haemoglobin that causes sickle-cell anaemia in certain African populations. However, the vast majority of genes are involved neither in the production of blood proteins nor in the production of pigments, and there was some concern that our emerging picture of evolutionary processes might be too heavily coloured (!) by the particular types of gene-product that were amenable to study.

Most genes, of course, make enzymes that catalyse the various pathways of metabolism, and until the 1960s very little was known of these genes – whether many of them were polymorphic at any one time, whether they were subject to rapid evolutionary change, and so on. The situation began to change in 1966, when the American geneticists Lewontin and Hubby applied the biochemical technique of electrophoresis to the detection of polymorphic enzyme variation in natural populations of *Drosophila*. These and very many subsequent studies revealed that a very large proportion of enzyme-producing genes are polymorphic at any one time in any natural population – not just in *Drosophila*, but in other invertebrates, in vertebrates and in plants.

Given what we already knew about *Biston* and some of the other 'classic' systems, the most obvious interpretation of the newer and more widespread molecular data was that all the newly discovered polymorphic loci were 'in transit' between an original monomorphic state and fixation of a new allele, originally introduced by mutation, which would then result in a second monomorphic state. However, various lines of argument rendered this unlikely. First, there are so many loci involved that a very large number of separate forms of natural selection would have to be acting simultaneously for this explanation to be correct. Second, even if this were the case, why should so many loci be *simultaneously* polymorphic? (Recall that in *Biston* populations going to fixation the polymorphic period was about 100 years; before that, there must have been thousands if not millions of years of phenotypic monomorphism in relation to the melanic/non-melanic locus.) Third, why should there be as much polymorphism in species inhabiting the relatively stable, deep-sea environment as there is in species inhabiting the most disturbed and transient of terrestrial habitats?

Because of the difficult of answering these and other questions, most evolutionary biologists were unable to believe that the widespread, newly found enzyme polymorphism was the same sort of phenomenon as the temporary polymorphic state observed in *Biston*. This does not mean that they all agreed upon some single alternative solution. Rather, views polarized into *two* alternative camps and a controversy arose that shook the whole foundation of Darwinism. In order to explain this controversy it is necessary to consider types of polymorphism not with respect to what sort of gene product is polymorphic but rather in relation to the *type of stability* characterizing the polymorphism. Hand-in-hand with this classification goes another – the classification of natural selection into different types, each of which is associated with one of the kinds of polymorphism. We now turn to these classifications, both of which hinge on the concept of stability.

Types of stability

It will help, since this subject is conceptually difficult, to focus upon an imaginary experimental situation. I envisage a population of fruit-flies (*Drosophila melanogaster*) established in an artificial environment such as a 'population cage' which allows a carrying capacity of about 1000 individuals, which can be maintained indefinitely through periodic renewal of the food-source. Consider a single locus making a single enzyme which is involved in the digestion of an important nutrient in the food-source. Let us assume that this locus is polymorphic with two alleles (and therefore three genotypes) and that the experimenter sets up his laboratory population with frequencies of alleles A_1 and A_2 both equal to 0.5. (He can achieve this by various combinations of genotypes. We shall assume he uses 250 A_1A_1's, 500 A_1A_2's, and 250 A_2A_2's, but we shall concentrate here on the *gene* frequency.)

Suppose our experimenter monitors the frequency of allele A_1 in his population over a period of ten or twenty generations. What will happen to it? The answer depends on whether there are fitness differences between flies of the three genotypes, and if so what form they take. We shall consider three distinct situations, as follows:

1. *Genotype A_1A_1 is fittest.* If the A_2 allele makes a defective version of the enzyme which is decidedly inferior at catalysing the breakdown of the main nutrient, then the A_2A_2 genotype will be the least fit, the heterozygote will be fitter (because half of its enzyme is of the more efficient sort) and the A_1A_1 genotype will be the fittest. In this situation, the frequency of allele A_1 will rise gradually every generation and, if the experiment lasts long enough, it will become fixed. This situation, which is a simplified version of what happened in urban populations of *Biston*, is referred to as *transient polymorphism*, and the form of selection 'driving' it is called, for obvious reasons, *directional selection*. The outcome of directional selection is independent of the starting frequency (see Figure 10*a*).

2. *Genotypes are equally fit.* If the only difference between the enzymes made by alleles A_1 and A_2 (which can be referred to as allozymes) is that they differ in one amino acid out of several hundred, with this particular amino acid being located outside of the active site and any other bits of the molecule which are crucial to its functioning, then it may be that the two versions of the enzyme are equally good at performing their task. In this situation, all three genotypes at locus A are equally fit. There is thus no force pushing the gene-frequency away from its starting

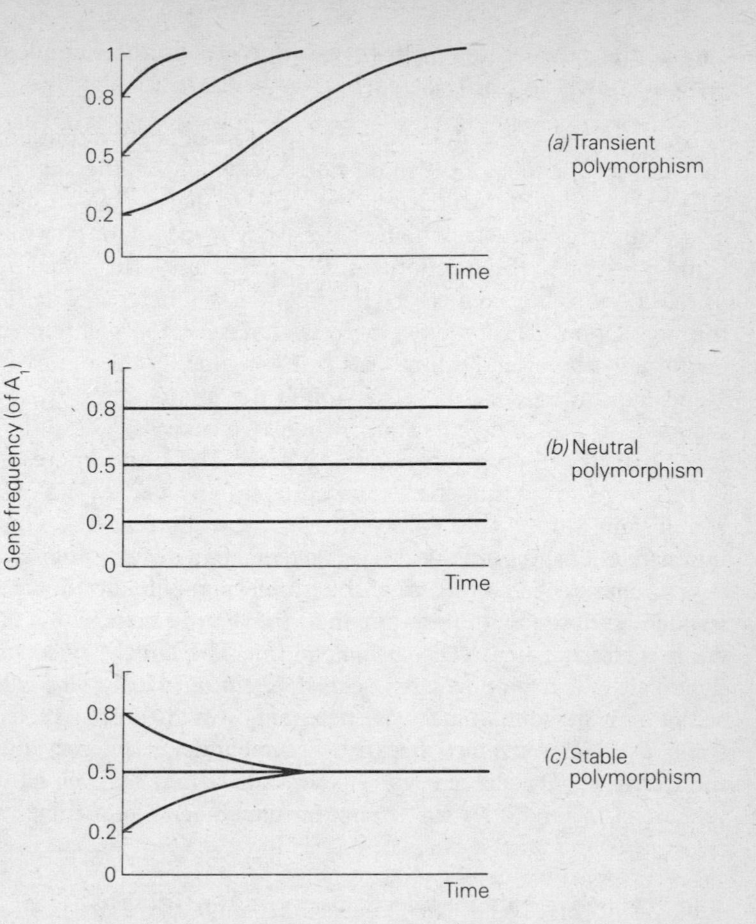

Figure 10. Kinds of polymorphism, distinguished according to their stability characteristics.

value, and consequently it tends to remain in the region of its initial frequency for the duration of the experiment. Unlike the previous situation, the final value is dependent on what numerical value is chosen for the initial frequency. If the experimenter chooses an initial frequency of 0.2 rather than 0.5, the actual frequency would remain in the 0.2 region (see Figure 10*b*). This situation is described as *neutral polymorphism* and it occurs when no selection of any kind is acting. The only process that can change the gene-frequency in this sort of situation is a random process known as *genetic drift.* This will be described shortly.

3. *Genotype A_1A_2 is fittest.* If our population's food-source contains two distinct versions of the main nutrient which require slightly different versions of the enzyme concerned to digest them, it may be that the heterozygote, which possesses two types of the enzyme, is the fittest of all three genotypes. We may picture this situation as selection 'trying' to increase the frequency of heterozygotes in the population. The problem here is that selection's 'intent' is continually thwarted because it is counteracted by the nature of Mendelian inheritance. However much the frequency of A_1A_2's increases in any one generation, when two heterozygotes interbreed half of their progeny are homozygotes. The long-term result of this 'conflict' is that the gene-frequency approaches a stable equilibrium value. This value, which, depending on the *exact* fitness relationships among the genotypes, may or may not be at 0.5, is approached regardless of the initial (non-equilibrial) gene-frequency chosen by the experimenter (see Figure 10*c*). This situation is referred to as *stable polymorphism.* The kind of selection described which produces it is called *heterozygous advantage* (or sometimes 'overdominance' or 'heterosis' – two terms which I think are better avoided because of ambiguity). Heterozygous advantage is only one of several types of *balancing selection* all of which act to produce a stable equilibrium in gene-frequency.

Allozyme polymorphism: selectionists versus neutralists

The great mass of allozymic polymorphism discovered in flies, men, plants, birds, snails and every other sort of organism that has been 'screened' using electrophoresis has to belong to one of the three categories just described. Of course, it is probably not true that these polymorphisms *all* fall into a single stability category. But most population geneticists believe that the great bulk of them (perhaps 90 per cent) do so. The question is: which category? As we have seen, there is no school of thought which proposes that the allozyme polymorphisms are largely transient. The two opposing schools that do exist are the *neutralists*, who reckon that the vast majority of these polymorphisms are selectively neutral and acted upon only by genetic drift (see below); and the *selectionists*, who believe that most of the allozymic polymorphisms are actively maintained by some form of balancing selection.

We have already seen how one form of balancing selection works, but we have not yet examined genetic drift, and now is clearly the time to do so. As its names implies, this process involves the gene-frequency drifting about in an 'aimless' way. In Figure 10*b*, where no selection was acting, the gene-frequency trajectories are depicted as perfect straight lines. This is obviously an idealized situation. In reality, animals are continually being born or dying in such a population, and this causes the gene-frequency to fluctuate slightly. If there is indeed no *systematic* force such as selection pushing it in one direction, the gene-frequency will tend to go up and down with about the same frequency. But this is a statistical process prone to doing, on some occasions, something different from what it 'normally tends to do'.

A good analogy to genetic drift is coin-tossing. The probability of getting a head in a single toss is the same as the probability that the gene-frequency will rise rather than fall in a single generation – that is, $\frac{1}{2}$. The probability of getting ten heads from ten tosses is the same as the probability of the gene-frequency going up rather than down in ten successive generations – $(\frac{1}{2})^{10}$, which works out at about 1/1000th. Thus if we have one experimental population monitored over ten generations, the chances of a continual rise in gene-frequency under genetic drift are very small; but if we have 10,000 populations we would expect about ten of them to exhibit such a rise (and another ten to show a similar fall). Drift will thus usually cause only minor fluctuations around a generally constant frequency, but will occasionally cause marked change and even fixation. The only difference between a natural population and our hypothetical experiment is that a new mutation comes into the population at a very low frequency rather than at 0.5. This means that it is much more likely to be lost than fixed through drift; but it still *can* be fixed, and it will go through a very long process of aimless wandering before it gets there. The neutralists believe that most observed enzyme polymorphisms are in just such a state.

Stalemate on present-day polymorphism

The obvious way to solve this controversy, it might at first seem, would be to set up experimental populations of the sort already described, and to see, for perhaps a hundred or so polymorphisms, whether gene-frequencies drift aimlessly or converge to a stable equilibrium. Unfortunately, there are many difficulties involved in attempting to do this. The two most serious are that (a) it is difficult to distinguish selection acting on one locus from selection acting at a neighbouring locus; and (b) at any rate if selection does act on enzyme polymorphisms it is almost certainly at an order of magnitude weaker than the selection observed in *Biston*, so experiments to detect it might take decades. Of the experiments that have been done, some have provided evidence for some form of selection and others have not; moreover, often these two alternative outcomes have been obtained from repeat experiments with the same polymorphism. As a result, no clear conclusion has emerged from these studies. However, there are signs that a conclusion may be emerging from a different sort of approach.

Molecular evolution

We can look at genetic variation in two quite different ways. In the preceding sections, the view taken was essentially an instantaneous one. That is, we look at the genetic structure of a population at one moment in time, or at least over a relatively short time-period. We thus observe the genetic variation in a way that is uninformative about the long-term evolutionary processes in which it may (or may not) be involved. The alternative approach is to compare a particular gene (or its product) between different species. Any differences that are found must have accumulated since the lineages leading to the two present-day species diverged. This sort of approach can now be applied in detail to both genes and their products because of our ability to determine the base-sequence of DNA and the amino-acid sequence of proteins.

Measurement of the degree of divergence has now been carried out for many interspecific comparisons and for many of the better-known proteins (haemoglobin, for example). Also, if we have a reasonable idea when the two lineages involved diverged from each other, it is possible to measure the *rate* of evolutionary change in the protein concerned per unit time. A large mass of data has now been built up on the evolutionary rates of proteins, and of the genes that produce them. What do these data tell us?

It is difficult to avoid drawing the conclusion that the data tell us that the neutralists were right. One of the main features of molecular evolution, as revealed over the last twenty years, is its low degree of rate-variability. When a particular protein is examined, its rate of evolution is quite constant, regardless of which evolutionary lineage it is in. This is what would be expected on the basis of genetic drift. If selection was the predominant cause of molecular evolution, we would expect that, as in *Biston betularia*, rapid evolution would accompany periods of marked environmental change, while periods of ecological constancy would be accompanied by much slower evolution.

At the DNA level, we again find a pattern of evolution that is readily compatible with the neutralist theory. If most new mutations accumulate simply because of genetic drift, they are (by definition) neither advantageous nor disadvantageous – they are merely *tolerated*, because they have no real effect. If this is the case, the fastest-evolving bits of DNA should be the least important ones, because the more crucial the function of a particular gene or part of a gene, the fewer the changes that will not be disadvantageous. As it happens, this is precisely what we find. In an 'ordinary' protein-producing gene, the redundant last bases in the triplet codes (see p. 41) evolve much faster than the other bases whose alteration causes a change in the structure of the corresponding protein. Also, in a class of non-functional genes called pseudogenes, *all* of the DNA evolves very rapidly, corresponding to the lack of protein effects of all bases in these regions of the genome.

It cannot be said that the selectionist theory is completely incompatible with the data that have accumulated on molecular evolution. Indeed, some geneticists still hold strongly to the selectionist view. But what can be said is that the neutralist theory assimilated the accumulating molecular data quite readily, while a selectionist interpretation seems to get more and more contorted as further data come in. Also, on an aesthetic level, the neutralist theory is preferable because it sees current polymorphism as a 'photograph' of continuing molecular evolution. Polymorphisms exist at one moment in time, and species diverge (at the molecular level) over long periods of time, because of one and the same process of genetic drift. In the selectionist view, directional selection drives evolutionary divergence, while the fundamentally different process of balancing selection causes the bulk of observed polymorphism existing at any moment in time. This is a much 'clumsier' picture. While aesthetic criteria are often dismissed by scientists as irrelevant to which of a pair of conflicting theories is correct, the philosopher Thomas Kuhn has painted a picture of scientific progress which sees criteria of this sort as very important in the replacement of one theory by another – and Kuhn's view has a ring of truth about it.

The neutralist theory as a Darwinian theory and as a selectionist theory

The evolution of selectively neutral characteristics through genetic drift has been referred to as non-Darwinian evolution, and the Japanese geneticist Motoo Kimura, who first put forward the neutralist theory, is seen by many as an arch anti-Darwinian figure. However, not only does Kimura acknowledge that morphological (as opposed to molecular) evolution is driven by natural selection; but indeed Darwin thought of 'polymorphism' as representing an exception to natural selection and a phenomenon that was effectively neutral (*The Origin of Species*, Chapter 4). Admittedly, Darwin was not referring to allozyme polymorphism, of which he was blissfully ignorant. Nevertheless, it is quite possible that, had this kind of polymorphism been discovered during his lifetime, he would have regarded it as neutral. I think that it is best to remain agnostic about which of the selectionist or neutralist theories is 'Darwinian'.

Quite apart from whether it is Darwinian or not, the neutralist theory is in many ways the ultimate selectionist theory, because it requires selection to have been very efficient at producing a near-optimal version of each enzyme shortly after it first appeared in evolution. If the vast majority of non-disadvantageous mutations are neutral rather than helpful, this would seem to imply that the molecule is near-optimal, for otherwise many mutational changes would make it function better. The neutralist view also implies that the optimal form of an enzyme is essentially independent of the external environment. For many enzymes, embedded deep within the metabolic network, this may well be the case. With the few that deal with external substrates, such independence is less easy to accept, and it may not be coincidental that the best evidence of selection at the allozyme level comes from just such enzymes – notably alcohol dehydrogenase in *Drosophila*, studied by the British geneticist Bryan Clarke.

So what?

While the selectionist–neutralist debate continues to cause heated controversy among population geneticists, other biologists often find this heatedness difficult to comprehend. Some are even unaware of the controversy. When told that there is a rival theory to Darwin's they typically become very interested, but when told what it is and that it only applies to molecular-level variation, a common response is 'so what?' To put it in a little more detail, biologists from other fields often view Kimura's theory as quite acceptable but not very revolutionary, in that it does not apply to the 'important' features of evolution, such as adaptation.

Interestingly, the two types of phenomenon for which alternative explanations to Darwinism (and its modern counterpart, 'neo-Darwinism') have been advanced are at the 'very small' and 'very big' ends of the scale of evolutionary events. The neutralists are essentially saying that there is a class of variant where the differences between rival forms are too small for selection to act. The other area of controversy (which will be covered later) is whether conventional neo-Darwinism can explain speciation and the origin of major taxonomic groups such as insects, molluscs and vertebrates. The two separate controversies involve doubt over different things. The three pillars of neo-Darwinism are selection, gradualism and a focus on differences in fitness between *individuals* rather than groups. The neutralists have questioned selection – albeit only for a particular class of variants. The other controversy involves questioning of the other two 'pillars', and, as we shall see, takes a very different form.

It is ironic that the 'middle-ground' between allozymes and speciation lends itself best to the least disputed and most conventional selectionist explanations. This ground is, as we have seen, represented to a large extent by studies of pigmentation – a type of character which Darwin regarded as a special case and which he described as the 'most fleeting of all characters'.

7 PARTICULATE INHERITANCE: THE INSIGHT OF MENDEL

The difficult thing to understand about Gregor Mendel's theory of heredity is not the mechanism involved, which in abstract form is remarkably simple, but the nature of the advance Mendel made over the previously prevailing view of inheritance. The problem arises because that earlier view has all but disappeared from the biological literature, and biologists today are exposed only to Mendel and to recent amplification of his ideas. Thus we are almost forced to equate Mendelism with inheritance. This is a bad starting point if we wish to get a clear view of Mendel's contribution; consequently, I need to begin this chapter by forcing the reader with some prior experience out of his comfy 'Mendelocentric' position. Readers with no biological background are actually at an advantage here, but if you fall into this group beware: it will be necessary to re-read the 'Basic genetics' section (pages 44–9) if you are to get the most out of the ensuing discussion.

To start with, then, we must put aside the gene-based terminology of the previous chapter, and consider, as Mendel did, organisms as entities containing mysterious 'factors' of a black-box nature which somehow codify information and are passed on to the progeny. A basic question, in this context, is – how many factors does each organism possess? It will be necessary, here, to consider only one characteristic of the organism at a time. I shall talk from now on about one such character – the size of the adult organism, measured in one dimension in some standard unit of length such as centimetres or inches. It does not matter much for the moment what our organism is. The height of a man or a plant, the diameter of a snail shell, and the length of a rat, are all cases of unidimensional size measurements. As has been mentioned, one of Mendel's experimental studies was on the inheritance of height in garden pea plants.

Our question about the number of factors can now be expressed in a more specific way. We may, for example, choose a particular species of plant, and ask: how many factors does each individual plant possess that represent its height in codified form and that pass on that information, or code, to the offspring? The simplest possible answer is clearly 'one', and this was widely believed to be the true answer prior to Mendel's work.

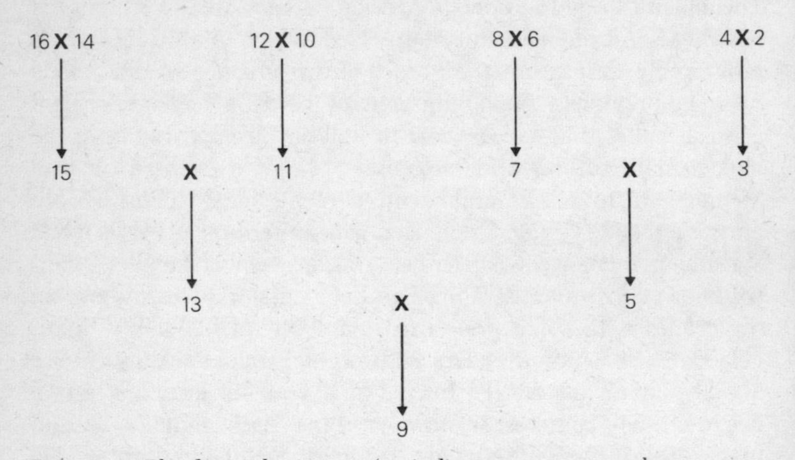

Figure 11. Blending inheritance. (× indicates a cross; numbers are plant heights.)

If each individual has only one factor representing height, then during reproduction the paternal and maternal factors must fuse or blend (hence the title 'blending inheritance') to give rise to the single factor present in each of the progeny. However, if this were to happen, any variation in height existing in a population of plants would rapidly be eliminated, leaving the population completely homogeneous. This result can easily be seen from the simple example given in Figure 11, where each plant has a single factor, labelled with a number representing the plant height (in centimetres or whatever) to which that factor gives rise.

What happens in the figure is that an initially wide range of variation in plant height is rapidly reduced, and we end up with all plants being 9 units tall, this being the mid-point of the original variability. Of course, the figure contains certain restrictions. For simplicity, only one offspring is shown for each cross. This is irrelevant, because in this system of inheritance all offspring would be homogeneous anyway. Another restriction is that the crosses shown involve each plant being paired with one that is adjacent to it in the height range. However, again, this restriction of the actual pattern shown is not necessary to cause the reduction and eventual elimination of the variability. Any pattern of crossing, other than the extreme case of the members of each pair always being of identical height, will have the same eventual result. It is the phenomenon of fusing or blending of the factors, which is at the base of this whole scheme, that causes the loss of the variation in height that was initially present.

The major problem with blending inheritance is that it is difficult to see how organisms employing this mechanism could evolve. Natural selection can act only when variation is present in a population, since it takes the form of some variants leaving more progeny than others. If the population is homogeneous, there are no variants, and the idea of selection becomes meaningless. In the population shown in Figure 11, this is the case from the fourth generation onwards. Of course, there will always be the occasional mutation, so an element of variation will be introduced now and then. But since mutations are known to be very rare events, and since their effects would anyhow be annulled shortly after their occurrence by the blending inheritance system, the amount of evolution that they would allow would be negligible.

Darwin was well aware that the then-believed system of inheritance was problematical for his theory of natural selection. Ironically, the solution to the problem was being worked out at about the time that *The Origin of Species* was published. Mendel published the results of his experiments in 1866, a mere seven years after the appearance of Darwin's famous volume. Unfortunately, though, even when Mendel's results became freely available, their significance was not appreciated until their so-called 'rediscovery' in 1900. Perhaps Mendel was, as they say, ahead of his time.

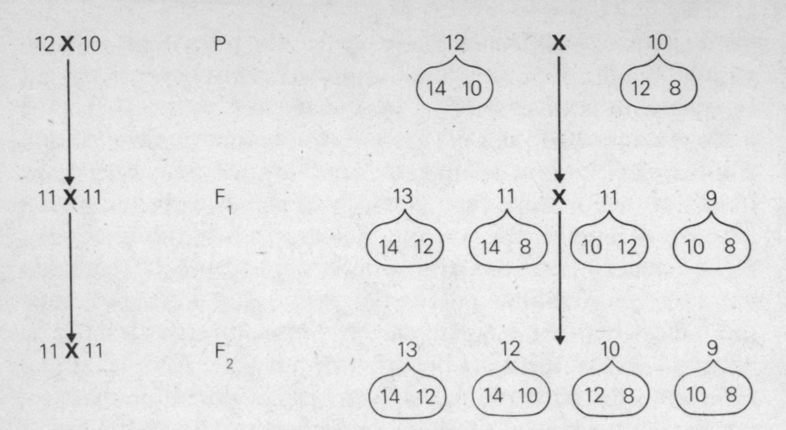

Figure 12. A comparison of blending (*left*) and particulate inheritance. (*Left*: numbers represent factors and phenotypes; *right*: encircled numbers represent factors, other numbers are phenotypes.)

The fundamental particle

What, then, was Mendel's penetrating insight that went unappreciated for more than thirty years and that eventually revolutionized the study of inheritance and gave birth to the science of genetics? Well, what he proposed was that for any characteristic, such as the height of a pea plant, each organism had not one factor but two. So far, this sounds like the most trivial of differences. However, the possession of two factors allows a fundamentally different kind of inheritance that is denied to organisms with only one. The factors can now behave as 'particles' which remain intact at fertilization rather than blending together. The combination of factors gives the organism its own phenotypic characteristics, but this combination is *not* passed on to the progeny. Rather, the factors separate out again, still behaving in a particulate manner, and contribute *independently* to the next generation.

This system can easily be contrasted with the previous one by looking at what happens in the cross 12 × 10 taken from Figure 11. On the left-hand side of Figure 12, this cross is shown again, still operating under a system of blending inheritance, but with two progeny shown rather than one, and with two further generations of inbreeding given. (This is a cross between two sibling organisms, not self-fertilization. The same experiment can be done with animals such as *Drosophila*.)

On the right-hand side of Figure 12 is the same starting point – plants with heights of 12 and 10 units. However, we now regard the actual height of any one plant, the phenotype, as being derived not from a single factor but from two. In the example given, the 12-unit plant has factors for 14 and 10 which remain separate for passing on to the next generation and only temporarily combine to give a phenotype of 12 in that one particular plant. A cross now takes the form of each plant giving *one* of its factors to each offspring.

As can be seen from the figure, there is no rapid elimination of the variability when inheritance operates in a particulate, as opposed to a blending, manner. In fact, although a diagram to show it would be fiendishly complex, the variation is not reduced *at all* with particulate inheritance; after hundreds of generations, you still have all the potential variation that you started with. There is thus no problem of lack of material upon which natural selection can act.

Formalizing the revolution

The picture of particulate inheritance given in Figure 12 is decidedly clumsy. It is also just an example rather than a general scheme. We can, however, solve both of these problems at once, as follows. Suppose the factor determining either the height of our plant, or any other characteristic of any sort of organism, is represented by A. Further, suppose that there are only two variant forms of this factor, A_1 and A_2. Then one possible mating between two individuals would be as shown below:

$$
\begin{array}{ccccc}
A_1A_1 & \times & A_2A_2 & & P \\
 & \downarrow & & & \\
 & A_1A_2 \times A_1A_2 & & & F_1 \\
 & \downarrow & & & \\
A_1A_1 \quad A_1A_2 & & A_1A_2 \quad A_2A_2 & & F_2
\end{array}
$$

This picture now connects with the basic genetics and associated terminology of Chapter 3. What we have done here, in terms of that terminology, is to cross two homozygotes, obtained an F_1 generation consisting of heterozygotes, and inbreed these to obtain a variable F_2 generation with a ratio of one A_1A_1 to two heterozygotes to one A_2A_2 – a well-known Mendelian ratio.

One thing that is not shown in the above formalization of Mendelian inheritance is the gamete stage. The arrow leads straight from one generation to the next. What is happening, although it has not been made explicit, is that in getting from P to F_1 generations, the A_1A_1 individual forms haploid A_1 germ-cells, and the A_2A_2 individual forms haploid A_2 germ-cells. To go from diploid germ-cell precursors to haploid germ-cells requires a separation, or segregation, of the two gene copies in the individual concerned (as we now know happens in meiosis; see p. 48–9). Mendel's first law, often called the law of segregation, simply asserts that this always happens. The second law, or law of independent assortment, states that if there is a second locus B coding for some other characteristic of the organism, then segregation at the A locus has no effect on segregation at the B locus, and *vice versa*. (Why the new word 'assortment' is brought in rather than referring simply to independent segregation I have no idea.) We now know that Mendel's first law is generally true and that the second is true providing the two loci concerned are not carried on the same chromosome.

Before leaving the subject of Mendel's laws, I would just like to re-emphasize that they are indeed *laws*. As already noted, scientific generalizations are usually only referred to as laws when they make predictions that can be confirmed *quantitatively*. Mendel's scheme of inheritance makes just such predictions: the 1:2:1 ratio mentioned above, and many others relating to different sorts of crosses. Using suitable experimental systems, all these predicted ratios can be, and have been, confirmed.

Back to blending?

There is something else missing from the formalized diagram of particulate inheritance given above, besides the germ-cells – namely the phenotype. Of course, the A's may refer to *any* characteristic since the formalization is a general one. But suppose for a moment that they refer again to height, whether of a pea plant or a man. What, then, are the phenotypes? We do not need to give actual numerical values here; we just need to consider the range of possibilities to appreciate the problem. With the A_1/A_2 system, there are three genotypes and so a maximum of three phenotypes. If there is dominance (say of A_1 over A_2), we are down to only two phenotypes. So we have tall, medium and short plants or, even worse, plants that can only be tall or short. In the real world, however, the character 'height' rarely behaves in this way. In our own species, adult height ranges continuously from less than five feet to more than six, and within the range of variation there are no distinct 'types', but rather a continuous range of finely intergrading values.

In fact, most characteristics of living organisms vary in a continuous way rather than consisting of a few discrete categories. Examples of the latter can indeed be found, such as the ABO blood group system in man, and also Mendel's peas, which were unusual in that they exhibited discrete variation in height (tall versus short). However, not only is height usually continuous in its variation, but so are weight, shape and many other characters. Does this mean that their inheritance obeys laws different from those of Mendel – perhaps even a system of blending inheritance?

The answer to this question is a categorical 'no'. Characters with continuous variation are underlain by a Mendelian form of inheritance. However, the link is a difficult one, and it was not fully understood until nearly twenty years after the 're-discovery' of Mendelism. The man who was most instrumental in forging this link was the British geneticist and statistician Sir Ronald Fisher.

I shall now attempt to explain, with the aid of Figure 13, how some characters are manifested as a continuous series of variants despite being produced by a particulate genetic system. Our starting point is the curve at the top of the figure (referred to as a normal or Gaussian curve) which is the pattern to which variation in height, weight and so on usually approximates. I have labelled the axes so as to refer to the distribution of adult shell diameter in a population of snails, and there is a reason for this choice: snails

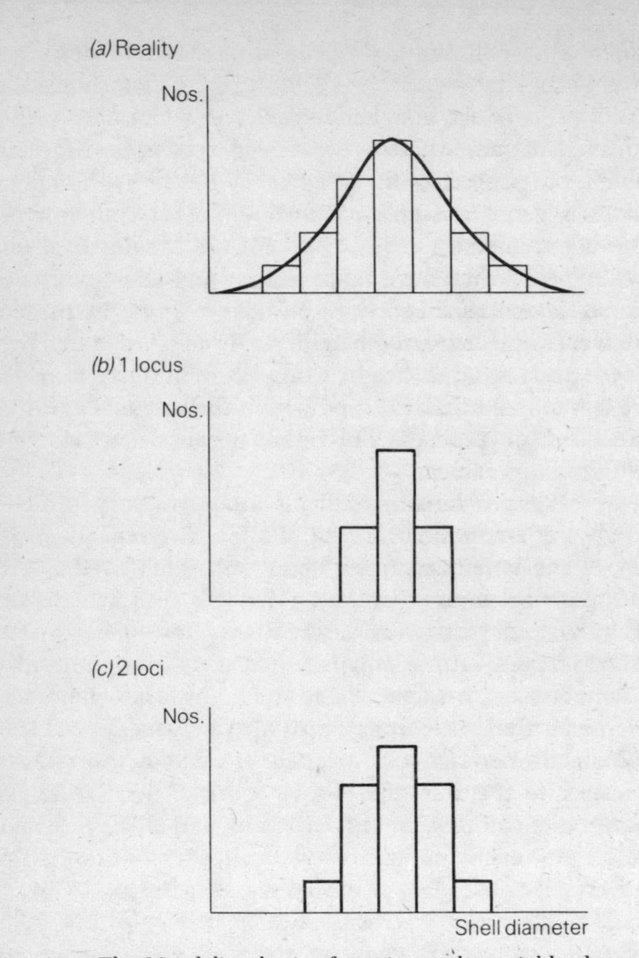

Figure 13. The Mendelian basis of continuously variable characters.

are hermaphrodite and we thus avoid the extra complication that exists in many organisms, including man, that males and females may show different distributions.

In any actual study, we take a representative sample of the population, measure each individual, and group them into size categories. The end-result of this procedure, which is also shown at the top of Figure 13, is a histogram which approximates the normal distribution. Both the curve and the histogram have two main features: they 'peak' in the middle, and they fall off symmetrically about this peak – we don't find a long 'tail' of the distribution on the left or the right.

Turning now to the genetic basis of the diameter of our hypothetical snail shell, let us first consider the simplest possibility – one locus with two alleles, and hence three genotypes. If there is no dominance, we also have three phenotypes, and let's suppose that the phenotypes corresponding to the three genotypes are as follows (measurements in millimetres):

A_1A_1 23
A_1A_2 22
A_2A_2 21

Another way of expressing this situation is to say that the A_1 allele adds a unit to diameter and the A_2 allele removes a unit. So an alternative to the previous display is:

A_1A_1 +1
A_1A_2 0
A_2A_2 −1

Now although a population in which the shell diameter is determined in this way has only three phenotypes, the distribution of snails among the three phenotypic categories is (providing alleles A_1 and A_2 are equally common) the same ratio that we met earlier in a simpler context – namely 1:2:1. (This numerical result, and extensions of it, is referred to as the Hardy–Weinberg equilibrium after its two co-discoverers.) A histogram showing this 1:2:1 distribution is shown in the middle of Figure 13. Although the range of variation is slight, we immediately notice that even this simple distribution possesses the two features noted earlier, namely a central peak and a symmetrical fall-off.

If we allow ourselves to consider the slightly more complex situation of two loci providing the genetic basis of shell diameter, with the second locus (B) having a similar effect to the first, namely

B_1B_1 +1
B_1B_2 0
B_2B_2 −1

then the distribution of phenotypes in the population gets closer to the picture that we started with. In the 2-locus case, we have a range of 'combined genotypes' with corresponding phenotypes, which are as follows (most intermediates omitted):

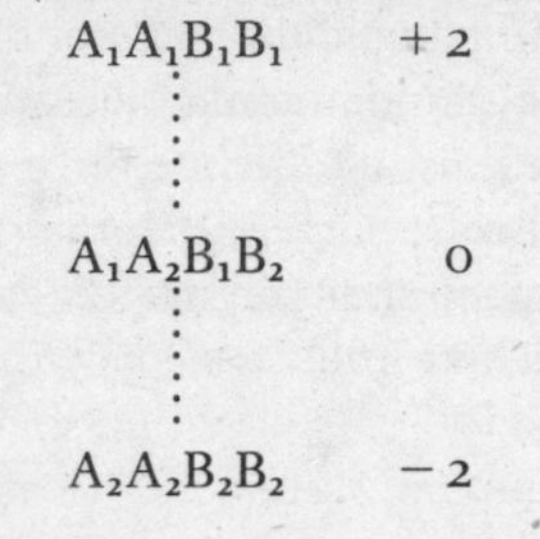

It can be shown, again by reference to Hardy–Weinberg equilibrium, that the distribution of shell diameter in the population now looks like the bottom histogram in Figure 13. Clearly, this is much more like reality (top picture) than the 1-locus case, since the range of variation is now wider.

As the number of loci is increased, the distribution of phenotypes gradually converges to the normal curve. Experiments with *Drosophila* and mice have shown that for most characters like size and weight the number of loci affecting them is large – usually in the range 10–100, though there are technical difficulties in producing precise estimates. Moreover, the distinction between adjacent phenotypic categories is blurred because this kind of character is affected by the environment (food, temperature, etc.) as well as by the genes. This is true of some characters, such as weight, more than others, such as height, but all characters show *some* environmental effect. Thus with many genes of individually small effect superimposed upon each other, and with environmental effects superimposed upon that, we end up with continuous variation in phenotype despite particulate inheritance.

Departures from Mendelian inheritance

The inheritance of continuously variable characters, as described in the previous section, represents an extension of, rather than a departure from, single-locus Mendelian theory. However, there are some cases in which a genuine departure from Mendelism does occur, and it is interesting to consider some of these briefly.

It will be clear, of course, that the basic Mendelian system, as described in the preceding sections, relates to sexually reproducing diploids. The way in which characters are inherited in organisms that are asexual and/or haploid is different. Nevertheless, in no case do we have such a fundamental difference that the system is not particulate. Genes always behave in a particulate manner; what varies is whether (and how) the genes of one organism combine with those of another (sexual systems) or not. Finally, it should be mentioned here that, even in organisms that usually reproduce asexually, it is often found that there is some occasional sexual form of propagation; and in such cases the normal Mendelian rules apply.

Even in sexually reproducing diploids, there are some characteristics that are inherited in a non-Mendelian way. In particular, I want to mention two different systems, both of which, confusingly, are sometimes referred to as 'maternal inheritance'. One of these involves traits that are coded for not by the main set of genes, borne on the chromosomes in the cell's nucleus, but by the small, subsidiary set of genes found in certain cell organelles, such as mitochondria. We usually find that sexual reproduction involves a male gamete that is little more than a mobile nucleus; and a female gamete that contains both nucleus and cytoplasm, and indeed typically much more cytoplasm than found in other cell types. Since each individual offspring is formed from one of each of these kinds of gamete, it follows that while it receives equal numbers of *nuclear* genes from its mother and father, *all* (or sometimes very nearly all) of its organelle genes are derived from its mother. It is for this reason that the term 'maternal inheritance' is sometimes used; but I think that 'organellar inheritance' or some other such label would be preferable for reasons that will become clear very shortly.

Another departure from the standard Mendelian picture occurs when a phenotypic character in a particular organism is determined not by that organism's own genome (nuclear or organellar), but rather by its mother's (nuclear) genome. An example of this is the phenomenon of chirality in snails. 'Chirality' refers to the direction of coiling of the shell. If we look down upon a snail's shell from above, we find that some are coiled in a right-handed (dextral) direction, while others are coiled in the opposite way (sinistral). Now in most species of snail, all the individuals are dextral, or else they are all sinistral. However, a few species exhibit variation in chirality, with some individuals being sinistral and others dextral, even in the same local population.

Geneticists interested in this phenomenon have used those species which contain both dextral and sinistral snails to set up crosses between the two types. The first experiments were performed on the pond snail *Lymnaea peregra*. The results appeared unintelligible until it was realized that the genes were behaving in a standard Mendelian manner, but the phenotype relating to a particular genotype was found in the next generation. The result is that any Mendelian phenomenon is found one generation later than would normally be expected. For example, the 1:2:1 ratio associated with the F_2 generation (which becomes 3:1 when there is dominance, as there is with chirality) was actually found in the F_3 generation in the *Lymnaea* experiments.

Although in this system the phenotype is determined by the mother, the genes involved are located in the nucleus and not in any organelle. The full details of the mechanism at work here are not yet clear, but the story so far goes something like this. The direction of coiling of a snail shell is determined not when the shell itself starts to form, fairly late in embryonic development, but rather when the fertilized egg undergoes its first few divisions – that is, at the very start of the developmental process. In a wide array of organisms, including such diverse forms as snails, flies, sea urchins and amphibians, there is evidence that, in very early development, the developing organism's own genome is not yet active. Nevertheless, cell divisions are occurring in an organized way, and so are clearly being controlled by something. That 'something' appears to take the form of information-bearing macromolecules (possibly long-lived RNA) deposited in the egg cytoplasm from maternal cells that were adjacent to the egg prior to its release from the mother. So of course these developmental 'messages' are coded for by the mother's genome because they are made in the mother's cells. Any character to which they give rise is also determined by the mother, regardless of what individual that character appears in.

I would like to make two final points on this interesting system.

First, for obvious reasons, it can be referred to as maternal inheritance; yet the mechanism involved is quite distinct from the other form of 'maternal inheritance' described earlier. So again I would advocate abandoning the term completely because of its ambiguity. A useful alternative label for the system seen in *Lymnaea* is 'delayed Mendelian inheritance'. Second, there is a link here with general theories of development. The existence of genes acting upon the next generation is not confined to chirality, nor for that matter to snails. Genes acting in this way are also known, for example, in *Drosophila*. A later chapter will show that some theories of development require inter-generation developmental messages, while others are more applicable to systems in which there are no such messages. As we have already seen, we can operate from the firm base of an accepted theory or mechanism in one area of biology to refute theories in conceptually linked areas. Perhaps, using what we know of developmental messages from genetic studies, we can begin to make a choice between alternative theories of development.

Genes and jumpers

The conventional Mendelian gene knows its place – and stays there. For example, the gene in *Drosophila melanogaster* which can mutate to give deformed 'vestigial' wings is on chromosome number 2, and is located about two-thirds of the way along that chromosome, at 'position 67'. This is true of different cells in the same fly, of different flies in the same population, and of populations taken from anywhere in the species' geographical range, which is essentially world-wide.

Until recently, it was thought that all genes behaved in this way, that is, that they were static entities with a fixed chromosomal location. However, it is now known that while this is indeed true of the majority, there is a minority of which it is not true. These genes jump, or to use the technical term transpose, from one position to another. Thus in some flies they are found at a certain chromosomal location (or locations, as they are also found in multiple copies) while in other flies their locations are different.

The piece of DNA that jumps does not correspond exactly to a gene. It may consist of several genes (one coding for an enzyme that causes transposition), or a single gene; or it may be too short to contain a functional gene at all. Whatever size it is, though, the mobile piece of DNA, referred to as a transposable element or transposon, can be recognized by certain distinctive DNA sequences at either end of it.

Transposons have been found wherever they have been looked for, including bacteria, yeasts and maize, as well as *Drosophila*. It is not clear whether they have a function, and if so what it is. They have been labelled 'selfish DNA' by some workers who see their only function as being their own survival and multiplication. (The act of transposing to another location is usually replicative – a copy is also left at the 'old' location.) It certainly is true that transposons can disrupt other genes or gene complexes by jumping into the middle of them. One intensive study of the molecular mechanisms of mutation in a gene complex of *Drosophila* found, surprisingly, that most mutations resulted from insertion of transposons.

The discovery of transposons poses all sorts of interesting questions about their function (if any) and their evolution. They also form a link in what is fast becoming a continuum from conventional static genes through transposons (which, as we have seen, move from place to place within the genome), then retroviruses (which move in and out of the genome and may cause some forms of cancer) to 'conventional' viruses such as those that cause the common cold. This sequence exhibits a gradual shift from entities that are functional parts of a larger organism to entities that are usually considered to be separate organisms in their own right.

Transposons and retroviruses are subversive elements in biological theory as well as in the organisms in which they are found. They extend the 'grey area' on the border of life that was mentioned in Chapter 2; they behave in a definitely non-Mendelian (albeit still particulate) manner; and they even come close to threatening Weismannism and its molecular counterpart the 'central dogma', as will be seen in the following chapter.

8 THE CENTRAL DOGMA OF MOLECULAR BIOLOGY

Weismann's view of the flow of information between germ line and soma was a unidirectional one. Information passes from germ-cells to the array of other cell types that make up the body of a multicellular plant or animal, but it never flows back from any of these cell types to the germ line. As we have seen, this view of the nature of information flow is compatible with Mendelism and Darwinism, but not with Lamarckism.

When Weismann advanced his views, during the second half of the last century, we knew next to nothing about the structure and function of living organisms at the molecular level. In contrast, we now have a mass of biological knowledge at this level, and new molecular facts appear almost daily. Among other things, we know which molecules carry genetic information, and which do not. It should therefore be possible to confirm or refute Weismann's abstract views on the basis of this recently acquired molecular knowledge.

To a rough approximation we can say that the body of an animal or plant is made of, and run by, proteins and their products. (Proteins form a major component of membranes, skin, muscle and hair, for example, and metabolism is governed by enzymic proteins.) Also to a rough approximation we can say that genetic information traversing the generations in the germ line is carried in DNA molecules. If we accept these statements as broadly true, then Weismannism can be rephrased something like this: information flows from DNA to protein, but not from protein to DNA.

This statement is in fact one of the main assertions of the 'central dogma' of molecular biology, first stated by Francis Crick in 1958. The other main assertion was that information encapsulated in nucleic acids can be perpetuated indefinitely, since nucleic acids can replicate; while the information contained in the amino-acid sequence of a protein is 'terminal', since a protein molecule cannot make another copy of itself.

The kinds of information transfer allowable under the central dogma can be summarized as follows*:

I shall concentrate here on the processes which involve transfer of information from one form to another rather than the maintenance of information in any one form by replication, since such *transfer* connects more readily with development (to be dealt with in the next two chapters), with Weismannism, and with evolution. This is not to say that the replication of information is unimportant, but merely that it is less essential in forging links between different areas of biological theory, which is one of the main tasks of this book.

Concentrating, then, on the information transfer side of the central dogma, the basic sequence is simply

DNA ⟶ mRNA ⟶ protein

The first step in establishing this sequence was of course Watson and Crick's elucidation of the structure of DNA in 1953.

The 1970s and 1980s have seen further major findings in molecular biology, one of which has caused us to modify the above picture. It has been discovered that in eukaryotes many genes are 'split', and have two or more informational or coding sequences (exons) separated by non-coding sequences (introns) which do not provide any genetic information that is converted into a protein sequence. What happens is that the whole gene, including introns, is transcribed into a large RNA molecule known as heterogeneous nuclear RNA, or hn RNA for short. It is heterogeneous because it represents both introns and exons; and it is called 'nuclear' because it never leaves the nucleus in that form. Instead, enzymes go to work on it while it is still within the nucleus, and 'chop out' the parts that correspond to the gene's introns. The exons are then joined together to form a shorter RNA molecule which forms the 'message' that will be translated into

* The dashed arrow indicates that RNA replication is a relatively rare process.

protein. So the process is one stage longer than we thought it was, namely

DNA ⟶ hn RNA ⟶ mRNA ⟶ protein

Actually, not all eukaryote genes are split, though the majority seem to be; and the genes of prokaryotes seem to be intron-free. So in some cases the simpler picture still applies, while in others the hn RNA stage is a necessary part of the process.

So far so good. Whichever of the two sequences from DNA to protein we consider, it is indeed a *unidirectional* sequence, as it should be if we are to have a molecular version of Weismannism. Also, we are not just talking, here, of studies on a particular organism, such as *Drosophila*, or even a particular Kingdom, such as Animalia. The sequence DNA to RNA to protein applies to plants, animals, fungi and bacteria, and indeed might at first appear to be universal in the living world. As usual, however, we run into difficulty in trying to make universal biological statements.

The departure from universality of the DNA to RNA to protein sequence occurs, predictably, in viruses. Specifically, it occurs in the group known as retroviruses. These are a sub-group of the 'RNA viruses' whose store of genetic information is carried in molecules of RNA rather than DNA. These viruses carry an enzyme (coded for by one of their own genes) called 'reverse transcriptase' which, as its name suggests, copies RNA into DNA. This enzyme violates the basic principle of unidirectional information flow that we have gradually been building. Actually, retroviruses go even further than just making DNA. This DNA is sometimes integrated into the host genome, and if this occurs in a germ-cell, the virus DNA is inherited by future host generations. At some stage in the future, the dormant virus DNA may suddenly start making more retroviruses. Thus these entities can be infective (like typical viruses) or inherited (like typical genes).

If we allow for the activities of retroviruses, and in particular the action of reverse transcriptase, to be taken into account, our diagram of information flow becomes

DNA ⇄ RNA ⟶ protein

Inspection of this flow pattern immediately raises the question: what about protein → RNA? In other words, is there an enzyme 'reverse translatase'?

The answer to this question is that no such enzyme has yet been discovered, and there is good reason to believe that reverse translation is not actually possible. This reason emerges from consideration of the 'genetic code' – the near universal way in which the amino acids that make up proteins are coded for by the nitrogenous bases that make up DNA and RNA. The important point about this code is that it is ambiguous in one direction but not in the other. Any 3-base RNA code specifies one and only one amino acid. For example, the code CAU specifies histidine. However, histidine does not imply a CAU code, since CAC also codes for histidine. So an RNA molecule giving a protein the information 'CAU' cannot be misinterpreted, but a reverse translation system giving the message 'histidine' in an attempt to build CAU into an RNA molecule would have a failure rate of 50 per cent. It is doubtful if a living system could cope with such a failure-prone system, and thus we have reason to suspect that reverse translation will never be found in nature. Whether this tentatively universal statement will go the same way as others in biology remains to be seen.

The existence of reverse transcriptase alone, then, poses no threat to Weismannism, because it only represents an incomplete and hence non-functional system for the 'backwards' flow of information from protein to DNA. If our Lamarckian giraffe's striving to reach high branches caused the formation of an enzyme which gave rise to more pronounced neck growth (and how even this could happen is not at all clear), there is no way in which this enzyme's structure could become codified in DNA and hence passed to future generations.

It is important to realize that the word 'information' as used earlier in this chapter means what one might call 'primary genetic information', that is, sequences of nucleic acid bases which give rise to sequences of amino acids in proteins. It is only to this sort of information that the central dogma, or Weismannism in general, applies. Proteins act on DNA in various ways, and indeed they actually *build* new DNA molecules when the process of DNA replication is taking place. Thus, in a broad sense, information of some sort is flowing from protein to DNA. Also, the germ line of an organism like *Drosophila*, while remaining distinct from the soma from an early stage in development, nevertheless receives all its nutrient molecules from the body, many of whose cell types have processed the food supply from ingestion, through several stages, before the germ-cells receive it. Again, information in a broad sense is flowing from soma to germ line, but this represents no threat to Weismannism, which relates only to what I have called primary genetic information.

One final comment is necessary on the link between Weismannism and the central dogma. The latter is frequently referred to simply as a molecular version of the former. However, as I pointed out earlier, this equating of the two theories is not strictly valid, since the central dogma has a 'replication' assertion as well as an assertion about information transfer. Even then, the information transfer component of the central dogma can be equated with Weismannism only if we treat the body as a bag of proteins and the germ-cells as packets of DNA; and both of these views are of course very simplistic. In particular, it is highly misleading to view the female gamete as a large DNA molecule. The egg cytoplasm is full of all sorts of agents, both nutritive and informational, which have an important role to play in the initial development of the new soma. Indeed, key processes of early development may be initiated by cytoplasmic agents rather than nuclear ones, as we saw with the gastropod chirality system in Chapter 7. I would even go further than this and say that informational molecules in the egg cytoplasm have a central position in any theory of cell differentiation or of development in general. Why I hold this view should become clear over the course of the next two chapters.

9 THE VARIABLE GENE ACTIVITY THEORY OF CELL DIFFERENTIATION

The problem of development, that is, the question of how a single cell (the fertilized egg) gives rise to an adult organism, is often referred to as 'the biggest unsolved problem in biology'. While this is indeed true of development considered in its entirety, we are beginning to make some headway in the understanding of certain components of the developmental process, notably cell differentiation. This phrase is used to denote the production of all the different cell types of the adult organism from, ultimately, the egg.

Cell differentiation is actually a very major component of development; and it is also a very dramatic one, as can readily be seen from a brief consideration of the number of different cell types and the diversity of their forms. In humans, for example, there are approximately 200 different types of cell. One well-known type is the red blood corpuscle (RBC for short), which is a bit like a miniature car-wheel. It has a fat outer periphery (equivalent to the tyre) and a thinner central disc (like the wheel itself). The diameter of one of these cells is less than a millionth of a metre, or, if you prefer, less than a thousandth of a millimetre. Contrast this with the length of a striped muscle cell such as would be found in your biceps: these can reach lengths of up to 40 mm. So from this comparison alone (and even more dramatic contrasts could be made), we have a differential in the cell's maximum dimension of about 40,000 ×. Accompanying this, of course, we also have a vast difference in shape. Needless to say, muscle cells are not 40 mm in all directions; rather, they are long and thin. With regard to their chemical components, these two cell types are again radically different. The RBC is packed with the oxygen-carrying protein haemoglobin, while muscle cells contain large quantities of the contractile proteins actin and myosin.

The RBC and the striped muscle cell are just two out of the many known human cell types. Others that are reasonably familiar are nerve cells, skin or epithelial cells, brain cells and white blood cells. Again, each has its characteristic size, shape and internal constituents – though I should perhaps admit that some of the types given above are rather broadly defined, and so are heterogeneous, having various sub-types within them.

These examples are taken from the human *adult*. Consider now a human being at a very early stage in his/her existence. Shortly after fertilization we have a solid ball of cells referred to as the morula. (This word comes from the Latin for mulberry; at this stage a human has a superficial resemblance to a berry of this general sort – raspberries and blackberries being better-known equivalents.) The cells of the morula, perhaps about fifty in number – though of course the number increases all the time – are to most intents and purposes identical. They are similar in phenotypic properties such as size and shape. Also, from a genotypic point of view, they all carry exactly the same set of genes, since they have all arisen through replication of the genome of the fertilized egg. What we have to explain, then, is how, during later stages of development, a bunch of very similar cells becomes a complex organism consisting of many dramatically different ones.

The two contenders

There are two fundamentally different ways in which diverse specialized cell types could arise from a single, more generalized cell. First, each specialist cell type could retain only that subset of its genetic material that it actually needs. Thus, for example, the genes for the muscle proteins actin and myosin would be lost in brain cells, RBCs and so on, and retained only in muscle cells, which themselves would have lost genes for haemoglobin and other proteins that they do not require. Depending on how genes for different kinds of proteins are grouped together on the chromosomes, this could involve the 'chopping-out' of lots of small bits of chromosome or alternatively the loss of whole chromosomes. Yet another alternative form of loss would be the permanent destruction of the unnecessary genes, but the retention of the irreversibly altered DNA in the genome.

In contrast to these various forms of 'selective loss' of genes is another possibility. Genes may be retained in the genome, and in a potentially functional form, but switched off by some molecular mechanism. This is called the 'variable gene activity theory' for the obvious reason that what it proposes is that it is the activity and not the existence of a gene that varies from cell type to cell type.

Several different kinds of evidence can be, and have been, brought to bear on this issue in attempts to distinguish between the selective loss and variable activity theories. However, I shall concentrate on just one experimental approach to this problem which has been particularly fruitful, and where the experiments themselves are especially fascinating, quite apart from their bearing on theory, which has been decisive. The type of experiment to which I refer is called nuclear transplantation. The major subject of these experiments has been the frog, and the experiments which I shall describe were performed at Cambridge by the British biologist John Gurdon and his associates.

Consider the life-cycle of a frog. It starts (if life-cycles can be described as starting anywhere) with the large eggs that make up 'frog spawn', which most of us will recall from childhood outings to ponds and streams. From the large egg develops a tadpole, and from that an adult. The whole process takes several months. Being a vertebrate, the frog is quite closely related to man, and has a not-too-dissimilar number of cell types, all of which arise from a bunch of very similar cells in early development and, ultimately, from the fertilized egg.

The rationale behind a nuclear transplantation experiment is as follows. First, we assume that the egg cytoplasm contains informational molecules that somehow start the whole process of development going. (Recall chirality in snails, where this is indeed the case.) What is to be tested is whether certain specialist cell types in the adult frog (or the tadpole) have specialized through permanent loss/destruction of a subset of their genes or through the (ultimately reversible) switching-off of such a subset. Now if it were possible to take the nucleus out of some one specialist adult cell and place this in nucleus-free egg cytoplasm, it might just be that agents within that cytoplasm would cause the development of many different cell types from the genetic information contained in the specialist cell's nucleus. This would be possible only if genes not necessary for the production of that specialist cell were switched off rather than being destroyed or lost.

Two things should be made clear about such an experimental design. First, it is in an important sense asymmetrical. A positive result will tell us that genes are merely switched off, and so will support the variable gene activity theory. However, a negative result will tell us nothing because we shall never be sure whether it was our experimental manipulation or the lack of the genes concerned that prevented the production of different cell types. Second, a positive result could take a variety of forms. Our 'hybrid' cell might produce a tumour-like mass of cells, containing many different cell types, or at best it might produce a perfect adult frog. There is no strong *a priori* reason to expect one or other of these outcomes.

The above point about negative results is the tip of an interesting philosophical iceberg. In practice, one hears a wide range of comments from scientists on this matter, varying from 'you can write off his experiments – the results were all negative' to 'a negative result is just as valid as a positive one'. This range of opinion reflects a range of truth; some negative results really are very informative, while others are of little value. Whereabouts we find ourselves on the spectrum from one of these extremes to the other depends entirely on the nature of the issue being tested and the nature of the experimental design.

Getting back to the specific experiments at hand, the first remarkable thing about them is that the techniques involved can actually work. That is, it is possible (1) to destroy an egg nucleus without destroying the cytoplasm, (2) to 'suck up' the nucleus of a specialist cell out of its own cytoplasm, using a micropipette, and (3) to inject that nucleus into the egg cytoplasm, giving a hybrid cell in the sense that the sources of its nucleus and cytoplasm are different.

What have these nuclear transplantation experiments produced? Well, the rather amazing answer is that after many initial trials it was eventually possible to obtain fully functional adult frogs. The donor nuclei were derived from two main sources – gut epithelial cells from tadpoles, and adult skin cells. The experiments thus proved beyond reasonable doubt that these specialist cell types contained all the genes necessary to make the full array of cell types that a frog requires. Thus, in the specialist cells from which these nuclei were taken, genes that were unnecessary in the functioning of – say – the skin cell were simply switched off, and were capable of being switched back on again by unidentified agents in the egg cytoplasm after transplantation.

We must, in fairness to the 'selective loss' theory, admit that Gurdon's results used only one particular sort of organism and only a few out of many cell types within that organism. One cannot proclaim the death of a theory in the light of what might be just a few exceptions to it – especially in biology, where, as we have seen, the general but not-quite-universal theory is the norm. However, various other lines of evidence have now been brought to bear on the struggle between the two contending theories of how cells differentiate, and with very few exceptions the variable gene activity theory is compatible with the evidence while the selective loss theory is not. It does not seem appropriate, in a book of this kind, to go into much detail on these other lines of evidence, and I shall restrict myself to brief mention of just two. First, there are a few cases where, in the natural course of development, as opposed to an experimental situation, a cell specializes in one way and then, at a later developmental stage, specializes in a different way. This phenomenon is known as transdifferentiation. Second, it is now possible to measure the amount of DNA in a nucleus, and it turns out that in most organisms the DNA content is the same across different cell types, at least within the range of experimental error. Of course, recent findings at the molecular level, particularly elements of DNA that jump into, around and out of the genome, make interpretation of data on amounts of DNA somewhat more difficult.

Mechanisms within theories

At one level, then, we have an answer to the question of how cells differentiate. They do so by switching on and off appropriate groups of genes, as proclaimed by the variable gene activity theory. However, that is not the end of the problem, for we would like to know *how* genes are actually switched on and off in the process of differentiation. If two major alternative ways in which this could be done were conceivable, then I might have entitled this section 'Theories within theories'. As it is, we don't have any two major alternatives – rather we just have a lot of molecular nuts and bolts out of which many putative mechanisms could be made, and which someone one day will use to put forward the correct mechanism. However, that day has not yet come, and at present we just have a few faint glimmerings of light in this area.

One thing that is fairly clear is that gene-switching usually takes place at the level of transcription rather than translation. Thus a gene that has been switched off is not making RNA which is somehow prevented from being translated into protein; rather, it is not making anything, RNA included. This seems reasonable in terms of natural selection acting in favour of cellular economy. A cell which wastes hard-won energy making useless RNA transcripts which merely clog up the nucleus and/or cytoplasm is likely to be an evolutionary disaster.

Another glimmer of light on the gene-switching problem in multicellular eukaryotes, with which this chapter has so far been exclusively concerned, is our reasonably advanced picture of gene-switching in bacteria. Nobody expects a mechanism as simple as the bacterial one to characterize the much more complex eukaryotic genome; but it is nevertheless helpful to consider the bacterial system, as it shows us what may be the simplest form of gene control. Using this starting point, we could then build models of increasingly complex processes until we eventually hit the right one. This is a rather idealistic picture of biological progress, but it does at least include the feature of 'minimal complexity' which is at the heart of all science: the simpler the terms in which a system can be described, the better.

The gene-control system known in bacteria is called the operon;

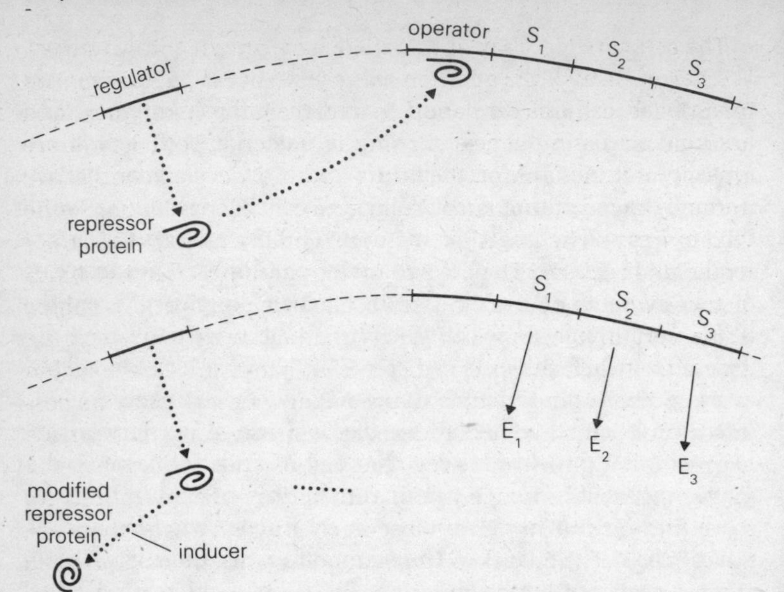

Figure 14. The operon system (based on the 'lac' operon). *Top*: Structural genes (S_1–S_3) switched off. *Bottom*: Structural genes switched on, producing enzymes (E_1–E_3).

a simplified diagram of the best-known example of this system (the 'lac' operon) is shown in Figure 14. What happens is that some structural genes (S1 to S3) which make enzymes may be prevented from transcribing by a regulator gene which makes a repressor protein. The repressor protein is capable of blocking the operator site, which must be unblocked if the structural genes are to work. However, if an inducer molecule enters the cell, it interacts with the repressor protein and modifies it in such a way that it can no longer block the operator, so the structural genes are switched on, and make their respective enzymes. If, as sometimes happens, the inducer is a nutritive molecule that is 'digested' by the enzymes E1 to E3, then it should be apparent that the system has homeostatic properties. The enzymes destroy the inducer, thereby causing their own genes to be switched off, but this switch-off means that the inducer is no longer destroyed, so further inducer molecules entering the cell will switch the genes back on again.

The astute reader may at this stage have grown a little uneasy. We started off with the question of the gene-switching mechanism involved in cell differentiation in multicellular eukaryotes, and are now discussing gene-switching in bacteria. Yet bacteria are typically unicellular, and they thus have no 'development' as we normally use the term, and certainly no cell differentiation. So the type of gene-switching that we were initially considering is not required of bacteria. The answer to this paradox is that two kinds of gene-switching occur in a multicellular eukaryote. First, there is the sort involved in cell differentiation, which involves the genes that produce 'specialist' proteins like actin, haemoglobin and keratin. Second, the genes producing 'general' proteins possessed by all cells and required for basic metabolic activities need to be switched on and off to regulate the amounts of these proteins in the cell. These two types of protein have been referred to, respectively, as luxury and housekeeping proteins, and these are useful labels. The eukaryotic regulation of housekeeping proteins may or may not have similarities to its equivalent in bacteria, namely the operon; and the switching on and off of genes for luxury proteins in eukaryotic cell differentiation may or may not be different from the 'housekeeping' form of regulation in eukaryotes. The weakest link of all, unfortunately, is between bacterial housekeeping regulation (the operon) about which we know the most, and gene regulation in eukaryotic cell differentiation, which is the most interesting problem.

Finally, a word of caution on the plethora of recent ideas on gene regulation in eukaryotes. As we have just seen, we don't yet know how this works. Also, there are many newly discovered kinds of DNA (introns, for example) for which there is as yet no known function. There is a temptation to 'kill two birds with one stone' and propose a regulatory role for one or more of these new types of DNA. I have recently seen regulatory models involving introns, and also a suggestion that pseudogenes are involved in regulation. (Pseudogenes are non-functional near-equivalents of certain functioning genes.) Personally, I don't believe either of these suggestions, and at this stage in the game I think it is best to treat all related attempts with scepticism unless accompanied by much persuasive data.

Renegade mechanisms

It is clear by now that loss of part of the genome from a cell is unusual, and that cell differentiation normally proceeds by gene switching. Consequently, the previous section was devoted to a discussion of mechanisms of gene-switching or, more accurately, what we don't know about them. However, there are certain situations in which some cell types do lose parts of the genome. In some of these there is no link with differentiation; in others there may be a partial link.

The most extreme case of loss of genetic material in a particular cell type occurs in the mammalian (including human) RBC. Here, the nucleus, and with it the entire genome, is lost, and the cell is left as just a bag of haemoglobin molecules. But this occurs *after* differentiation. That is, an appropriate pattern of gene-switching produces lots of haemoglobin, and few or no other proteins; *then* the nucleus is lost. Loss of the entire genome of course could not be a *cause* of cell differentiation, as it would lead to a wholly new cell type – the zero-protein (and hence non-existent) cell!

Less extreme cases where only part of the genome is lost are also known, and these are clearly better candidates for a role in cell differentiation. However, even where the loss is partial, and where it varies between tissues, it never seems to amount to a full-blown differentiation mechanism. For example in nematode worms part of the chromosomes are lost in somatic tissues but not in the germ-cells; but the same parts seem to be lost in different somatic cell types, which is a fundamental departure from the pattern that a differentiation-causing mechanism would reveal.

The sparsity of cases of gene loss, coupled with the odd patterns of loss that do occur, as described above, actually help to confirm rather than refute the variable gene activity theory. So in a sense we do have a general theory of cell differentiation, albeit the molecular mechanisms of eukaryotic gene-switching remain to be elucidated. However, there is one major problem with our theory, which can best be posed as the question 'where does it all start?'

The problem of beginnings

Suppose that in eukaryotes there is some molecular mechanism different in detail but similar in principle to the bacterial operon. Further suppose that this mechanism is involved in the gene-switching of cell differentiation as well as that relating to 'house-keeping' (which we need not consider here). Whereas in bacteria any inducer molecule entering the cell must necessarily come from the external environment, its equivalent in the eukaryotic system must enter from an adjoining cell, because any cell upon which we focus our attention is now embedded in a matrix of others. This in itself is not a problem; many small and medium-size molecules diffuse from cell to cell, though large macro-molecules like proteins and nucleic acids do not. Of course, the incoming molecule is unlikely to be nutritive (like the sugar lactose, which acts as the inducer in the famous 'lac' operon) if its function is to act as a trigger in cell differentiation. A more likely candidate for such a trigger would be an incoming hormone molecule – indeed hormones are known to have many developmental effects in a wide range of organisms.

The 'problem of beginnings' can now be put in the following way. If, in order to start a particular group of cells differentiating along a particular course, we need an infusion of hormone, this implies that there is already a specialist cell type whose function is hormone production. How did this cell type itself differentiate? Well, of course it may have done so in response to a different hormone – but that merely pushes the same problem back one stage, while leaving it intact.

I am going to leave this question hanging for the moment, because it is broadly applicable in the developmental sphere rather than applying only to cell differentiation; and it will thus be more appropriately dealt with in the following chapter. However, a brief nudge in the direction of a solution may be helpful here. It is in fact inaccurate to think of a fertilized egg as a totally un-specialized cell, and even worse to think of it as a homogeneous one. The cytoplasm contains many developmentally active agents, and these are often unevenly distributed. So when the egg begins to divide, some of its daughter-cells will contain agents that others do not. In a sense, therefore, there is no stage in the life-cycle that is entirely devoid of 'differentiation', although it cannot be *among-cell* differentiation at the egg stage, and it is not *externally visible* at slightly later stages such as the morula. Perhaps the problem we have to address in development is not how cell differentiation (and pattern formation: see next chapter) originates, but rather how certain kinds and amounts of differentiation and pattern formation lead to other kinds and amounts.

10 THE MISSING THEORY OF DEVELOPMENT

To my mind, development is the most amazing of all biological processes. There is something almost miraculous about the ability of fertilized eggs to become adult organisms. Indeed, I have often wondered why those who attack scientific theories of life (for instance the creationists, whom we shall deal with later) have concentrated so much upon evolution and largely avoided attacking our current mechanistic approach to development.

Consider for a moment what actually happens in the development of a multicellular creature such as *Drosophila* or man. At the beginning, we have a single cell, the fertilized egg. At the end, we have a vast number of cells – about a million in a *Drosophila*, many more in a man. The most striking thing, though, is not the increase in cell number, but the incredible constancy of the end-product. The vast majority of individuals of *Homo sapiens* or of *Drosophila melanogaster* end up falling within a very narrow range of variation. Admittedly, this variation is important for evolution, as we have seen. However, the slight variation in height, weight, shape and so on that we see around us in man, and its equivalent in other species, is trivial compared to the *potential* variation among entities composed of many cells. Think of the number of possible three-dimensional layouts that could be adopted by different 'humans' supposing that the only constraint was that there had to be the correct number of cells belonging to all the different human cell types, some of which (such as muscle cells and RBCs) we briefly looked at in the previous chapter. One type of 'human' would be a linear array of cells – perhaps starting at one end with so many muscle cells, then so many nerve cells, etc. – until all the requisite numbers of the requisite cell types were exhausted. If standing vertically, this 'linear human' would have his head literally in the clouds but would be too thin to see. You can easily invent all sorts of other humans – spherical ones with concentric shells each made up of a particular cell type, cubic ones, 'humans' resembling cars or lamp-posts – the list is endless.

Now this may seem like a frivolous exercise, but it is emphatically not; for it identifies the main problem of development over and above that of cell differentiation – how does the developing organism maintain the correct spatial arrangement of cells, for the species to which it belongs, at every stage of development and, ultimately, in the adult? This central problem is often referred to as pattern formation or morphogenesis. (Some authors try to split hairs and make a distinction between these two terms, but I shall use them interchangeably here.)

Of course, it is not by magic that humans conform closely to a certain basic structural pattern, and the same applies to all other species. Rather, the reason why your (adult) friends are all visible and between four and seven feet tall rather than being invisible skyscrapers or other bizarre forms lies in the way development works. Now this last sentence is uninformative, because we do not know how development does work or, more specifically, what it is that causes repeatable pattern formation – hence the title of this chapter and the fact, noted in the last one, that development is often described as the biggest unsolved problem in biology.

Our ignorance, however, is not complete. One thing that is clear is that the mechanisms waiting to be revealed are quite distinct from those of cell differentiation. What we are concerned with now is not the internal cellular changes causing a particular cell to become, for example, a muscle cell. Rather, we are concerned with what happens in *large populations of cells.* In other words, we have moved from the within-cell to the among-cell component of development. As we have seen, even when we assume that the correct number of cells have differentiated in the right ways, this still leaves endless possibilities in the sphere of pattern formation. Clearly, an additional process over and above the cell differentiation mechanism needs to be invoked, although the two must interact.

I would guess that at some level or other it will in the future be possible to provide a general theory of pattern formation. This is an act of faith more than anything else, though it is based to some extent on the existence of precedents in related areas. If we can have general theories of inheritance and evolution, why not one of development? In fact, general theories of development have been proposed in the past, and it may even be that we have a modern theory (or at least an outline theory) of development which is not widely recognized as such. These issues will be examined in subsequent sections, starting with an early theoretical controversy on the basic nature of development.

Preformation versus epigenesis

If, within the fertilized egg, there existed a perfectly formed but miniaturized adult organism, there would be no problem of pattern formation (or cell differentiation), only one of growth. This possibility, which today seems decidedly weird, was widely believed in the seventeenth and eighteenth centuries. It is known, for obvious reasons, as the theory of preformation. Belief in such a theory creates all sorts of difficulties, one of which is that you have to believe that a large number of generations (presumably all between the origin and extinction of that species) have to be 'telescoped' inside each other. That is, within the tiny man or 'homunculus' in the fertilized human egg is another one, and within that yet another, and so on. Another difficulty is that, given that the fertilized egg arose from the union of an ovum and a sperm cell, presumably only one of these contained the preformed adult. In fact, the preformationist camp was split into two opposing factions on this issue, with the 'ovists' believing that the preformed adult resided in the (unfertilized) egg and the 'animalculists' believing it to be in the sperm cell (or animalcule, as it used to be called).

The dispute between animalculists and ovists was 'solved' in a way that is interesting for its general philosophical message. The apparently decisive information came in the form of the discovery of parthenogenesis. This is a form of reproduction in which a female reproduces asexually, via unfertilized eggs. The existence of parthenogenesis seemed to prove the ovists to be right, since it was perfectly compatible with their views and incompatible with those of the animalculists. However, we now know that the overall theory of preformation was wrong, and therefore *both* of the major schools of thought within it were wrong. The compatibility of one of these schools with certain facts, such as parthenogenesis, is best described as coincidental.

There is a general point at stake here. I have previously stated that compatibility of two (or more) theories with each other does not necessarily mean they are correct – rather, both (or all) may be wrong, as in the case of blending inheritance/creation (see Chapter 4). I said then that in addition to demanding compatibility with each other, we must also demand that our theories link in satisfactorily with the real world – that is, they must be compatible with whatever relevant observational and experimental data are at hand. However, what we see now is that even this kind of compatibility is not sufficient to ensure that the theory is correct, because there may be other, conflicting theories that also appear to 'explain' the same data. If the theories are mutually exclusive, as is often the case, only one of them can be correct. How, though, do we choose the correct theory, if both or all explain the same data? There is no universal answer to this question, though it will sometimes be possible to distinguish the correct theory as the one which explains the widest range of observations. Newtonian and Einsteinian physics are a case in point: both explain physical phenomena under terrestrial conditions, but only Einstein's physics continues to give an adequate view of things under other conditions, such as speeds approaching that of light.

Getting back to the main argument here, if the preformationist theory was wrong, what was the alternative theory that was right? It was in fact the theory of epigenesis, proposed by the German embryologist Caspar Friedrich Wolff in the second half of the eighteenth century. According to this theory there was, as we now know to be true, *no* preformed adult organism within the fertilized egg; there was, rather, a set of instructions from which that adult would be built. In a sense epigenesis is a correct theory of development, though we rarely think of it as such because it now seems obvious that a preformationist view is unacceptable. This is not, though, to detract from the advance that Wolff made over the preformationists. But what Wolff's theory does not tell us is *how* the set of instructions in the egg becomes transformed into the actuality of the adult organism.

Genes versus cytoplasm

It is important to be clear about the exact nature of the controversy between supporters of the preformation and epigenesis theories. I say this not just because it is nice to have a firm grasp of the history of science, but because a clear understanding of this old conflict makes it easier to grasp a new and different one which I shall shortly describe. Basically, what was at issue was whether the future adult organism was present in *actual* or *codified* form in the fertilized egg. Clearly, it must be present in one form or the other. If there is no miniature organism to grow into an adult, or any code or set of instructions from which an adult can be built, the fertilized egg cannot produce an adult organism at all, unless we resort to vitalism of some kind.

The controversy between preformationists and the supporters of epigenesis was settled in an unusually clear and, dare I say it, universal way. No newly fertilized egg contains within it a miniature adult organism. Whether we are dealing with men, frogs, snails, flies or flowering plants, this statement is true without qualification. Nor, for that matter, is any compromise theory necessary in that there are not even *parts* of the adult organism present in the egg. We do not find miniature flower petals, insect wings, or anything comparable inside the eggs of the appropriate species. Thus *all* of the adult organism is present in codified rather than actual form in *all* fertilized eggs in *all* species.

This is an important conclusion because it naturally leads to the key issue of *how* the adult organism, and indeed all the developmental stages leading to it, are coded for in the egg. Now we have to be very careful here, because there are codes, or instructions (which I think is a preferable term), of two fundamentally different sorts. First, there are the instructions contained within the genome – the sequences of DNA that code for RNAs and proteins. With regard to this set of instructions – the genetic set – the *whole* adult organism is in a sense encoded in the fertilized egg. This follows from the variable gene activity theory of cell differentiation, which we examined in the previous chapter. Indeed, it follows from this theory that all diploid cell types of a multicellular organism (including, of course, the fertilized egg cell) contain the whole organism in codified form.

The second set of instructions, the importance of which is often unappreciated, is the cytoplasmic set, that is, the collection of informational (as opposed to nutritive or routine metabolic) molecules in the egg cytoplasm, and, later, in the cytoplasm of other cells. One reason for the relative neglect of these cytoplasmic factors is the current gene-centred approach in biology, which tends to see the genome as all-important and other 'merely phenotypic' entities as being in a sense subservient to it. In the developmental context, this gene-centred view is best seen in phrases such as 'the genetic control of development' and 'the genetic basis of development', the latter actually being the title of a recent book* – quite a good one as it happens.

What seems to me objectionable about these phrases is not that they are in some sense 'wrong'. After all, there is a genetic basis to development, and genes must, in some way, control developmental processses. Rather, I think they are misleading, because they concentrate too much on one side of a decidedly two-way mechanism. What happens in development is that cytoplasmic factors somehow switch on (or off) certain genes, which then make (often cytoplasmic) products, some of which act to switch on or off other genes or gene-sets, and so on. That is, there is a constant interplay between the genes themselves and non-genetic factors, and it is this interplay that produces the phenomenon we recognize as development. One side of the process involves the control of genes by non-genetic (or 'cytoplasmic', in a rather loose sense) factors; the other involves the control of these factors by the genes.

It is easy, in science, to fall into the trap of criticizing one's opponents, that is the supporters or apparent supporters of a view alternative to one's own, for blindly pursuing their own approach and for failing to see the obvious. This is not a helpful stance, because it sees one's associates as intelligent and one's opponents as stupid. Life is rarely as simple as that! I make this point here because I want to stress that most practising biologists who use a term such as 'the genetic control of development' would readily

* By A. D. Stewart and D. Hunt, published by Blackie.

recognize the other side of the coin. My objection is not to what is going on inside the head of one such biologist but rather to the 'atmosphere' that a group of such people unwittingly creates, which tends to push the collective impression of a process in a particular direction (in this case the developmental process in a genetic direction) without anyone intentionally setting out to do so.

Having gone to some pains to point out the importance of cytoplasmic factors in development, which will be wasted effort to those familiar with the work of C. H. Waddington, who made this same point very clearly many years ago, I now want to focus on whether the cytoplasmic set of instructions in the fertilized egg is 'complete', like its genetic counterpart. As we have seen, all genes are present in the fertilized egg, like any other diploid cell, and since all adult cell types use a subset of the full gene complement, the genetic basis of all adult cells is present in the egg. In this sense at least the adult organism is entirely codified in the single set of *genetic* instructions that comes across from the previous generation. What of the cytoplasmic set? Well, clearly, the full organismic complement of proteins is not found in the fertilized egg's cytoplasm; and in this sense the cytoplasm bridging the generation gap (which as I stressed earlier comes largely from the mother) is certainly incomplete.

But that is barking up the wrong tree. What is of interest in a developmental context is not all cytoplasmic factors but only those with an ability to influence the developmental process, that is those capable, directly or indirectly, of switching genes on or off. Anything *with* such an ability can be described as being a morphogen or as having a morphogenetic effect. Examples of morphogens include many hormones, but also lots of other kinds of substance, a good few of which probably remain to have their morphogenetic activity discovered.

Given, then, that the fertilized egg cytoplasm contains, among other things, a set of morphogens (whatever these may be in chemical terms), we may ask whether *this* set is complete. In other words, are all the morphogens required at all stages of the developmental process present in the cytoplasm of this single initial cell?

I don't think enough is known to give a definite answer to this question, but I suspect very strongly that the answer is 'no' and that only a small fraction of the morphogens required throughout development are present in the cytoplasm of the fertilized egg. Indeed, this view is implicit in the picture of development just painted, namely a two-way gene/morphogen control process. Some of the genes turned on by an early morphogen may act to produce other morphogens which were not previously present, and it is therefore unnecessary to have all morphogens present at the outset.

The rationale behind this focus on morphogens, within the cytoplasmic set of instructions, can also be applied to the genetic set. That is, we can focus not on the whole genome but rather on those genes – probably a small subset of the full complement – which act to govern aspects of the developmental process. These could be called morphogen-genes, but I have referred to them in another book (see Suggestions for Further Reading) as D-genes – short for developmental genes – and since this is a more compact label I shall continue to use it.

It is important to distinguish D-genes, which control development, from those other genes which *contribute* to development by being switched on in some cells but which themselves have no causal role in the process. For example, suppose we take a hypothetical D-gene which controls the differentiation of some cell types. As a result of this D-gene's activities, one group of cells may become muscle cells which contain, among other things, a protein called myoglobin; while another group of cells may become nerve cells, which do not contain this protein. The gene that produces myoglobin has contributed to cell differentiation in this example, but it was not one of the D-genes that caused the two groups of cells to begin differentiating in their respective ways. When we are dealing with pattern formation rather than cell differentiation the same distinction applies.

As with morphogens, we can ask, in relation to D-genes, whether all that are required for the full developmental process (that is, all of them!) are present in the fertilized egg. However, we need really to ask whether all D-genes are *active* at this stage rather than whether they are present, which we have already established is true of all genes including D-genes.

The surprising answer to this question seems to be that *none* of them is active. Evidence suggesting this has come in a variety of forms. Experiments with amphibians, in which all of the genetic material of the developing organism was destroyed, failed to prevent formation of the early (but multicellular) developmental stage called the blastula. Also, in *Drosophila* it appears that no products of the developing organism's own genes are found until the blastoderm stage of development (also early but multicellular). What this means is that the earliest developmental processes are under the control of morphogens in the egg cytoplasm rather than the genes. Although these early stages of development are much simpler than later ones, they clearly represent organized pattern formation. The blastoderm of *Drosophila*, for example, is a hollow, egg-shaped envelope of cells. (Initially, it is syncytial, or acellular, but later cell membranes are constructed.) It never takes any other form – linear, spherical or whatever. Thus it is clear that the initiation and earliest stages of pattern formation are not controlled by the developing organism's own D-genes but rather by cytoplasmic morphogens, themselves products of the mother's D-genes. The case of chirality in snails, mentioned in Chapter 7, is another example of this.

The morphogenetic tree

In the previous section I concentrated on what happens at the beginning of development, and came to the conclusion that only a *small subset* of morphogens and *no* D-genes were present in active form at that initial stage. I now want to consider development in totality, that is, the whole egg-to-adult sequence in an organism such as a snail or mammal, and to ask the question: how does the number of morphogens, and the number of active D-genes, alter as we progress through the sequence of developmental events?

Again, we do not have enough experimental data to give a categorical answer. However, I would like to suggest that the number of morphogens and/or the number of active D-genes *increases* during development, at least up to a certain stage. The reason for this conclusion is a simple one. Consider the development of a human, or other mammal. At a very early stage in development (at the blastula stage, for instance), the organism is developing as an integrated whole. Later on, different parts of the organism develop along quite distinct pathways. An example of such a 'quasi-independent part' is the mammalian limb bud, which eventually becomes an adult limb. Still later in development, many more parts are going their own ways. For example, your little finger and your index finger (neither of which are discernible at the earliest stage of the limb bud) become different in size and shape. Thus, as development proceeds, the developing organism essentially fragments into more and more parts, each making up a smaller and smaller fraction of the organism, and each having a fate (in terms of pattern formation) that is different from the others.

This very basic observation about development seems to suggest that the number of active D-genes and/or morphogens (I suspect both) increases throughout development, or at least up until the point at which most pattern formation is complete. This may actually occur quite early in development, for example after a few months in humans, to be followed by a period of simple growth. (This, incidentally, suggests that growth is yet another semi-distinct developmental problem, like pattern formation and cell differentiation.) This view of development is given in Figure 15. It needs to be stressed that I do not claim that the number of morphogens in any particular *cell* at an intermediate stage of development is greater than in the fertilized egg – indeed the reverse may well be true. Rather, the contrast that is being made is between the *whole organism* at two stages in development, though of course at the beginning it consists of only one cell.

I have called this picture of development the morphogenetic tree; but its tree-like or hierarchical nature is not perhaps made very explicit in Figure 15. Consequently, it will be useful to present the same view in a different pictorial way. This will be done in the next section.

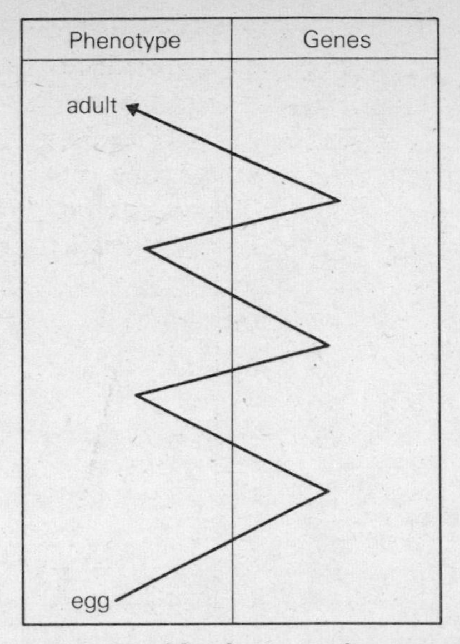

Figure 15. A schematic view of development, showing intercausality between genes and phenotype. The number of active morphogens and/or D-genes increases going 'upwards' from egg to adult. When the arrow crosses from left to right, this indicates phenotypic agents switching genes on or off. As it crosses back, this indicates genes influencing the construction of the phenotype.

Tree versus flag (or hierarchy versus line)

I want to introduce, at this stage, the concept of a 'heterogeneity'. This should not be too difficult – everyone is aware that homogeneous means things being the same in some sense while heterogeneous means things being different. In a developmental context, one sort of thing (but by no means the only one) that can be homogeneous or heterogeneous is a group of cells. If the cells are all the same, the group is homogeneous; if some are nerve cells and some are muscle cells, the group is heterogeneous. Providing that the overall group that contains the different cell types in the latter case is itself reasonably well defined, we may refer to it as 'a heterogeneity' or 'a system of heterogeneity'.

The important thing about heterogeneities in development is

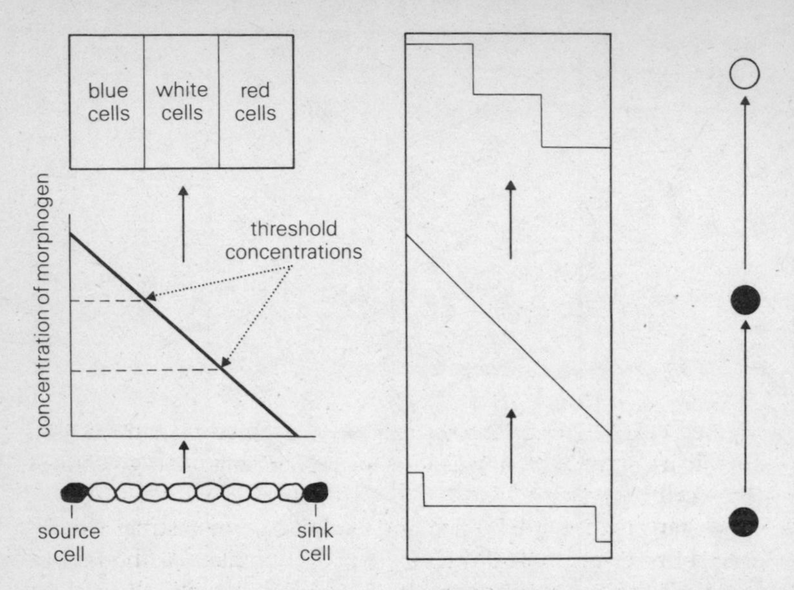

Figure 16. Three versions of the French flag model. *Left*: The 'complete' version. *Centre*: Abstraction revealing only the form of the heterogeneity. *Right*: Abstraction revealing whether each heterogeneity is morphogenetic or terminal (solid circle = morphogenetic; open circle = terminal). Arrows indicate causal links.

that they convey information while a homogeneity does not. One way in which a heterogeneity can give rise to information that is useful in pattern formation is illustrated by the 'French flag model', first put forward by the British embryologist Louis Wolpert, and shown on the left in Figure 16. What happens here is that a heterogeneity in the form of three cell types, one of which produces and one of which destroys a diffusible morphogen (the third being inert), gives rise to a second heterogeneity in the form of a concentration gradient of the morphogen. (This is an example of a non-cellular heterogeneity; its corresponding homogeneity would be a constant concentration.) If the cells of a developing sheet of tissue exposed to this gradient contain internal genetic 'switches' such that they differentiate in different ways depending on the concentration of morphogen, we end up with a spatially

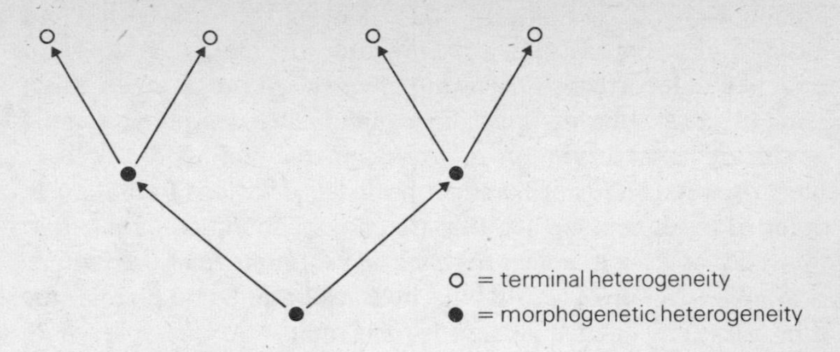

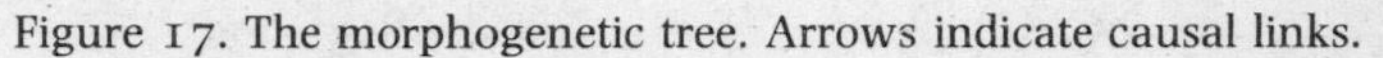
Figure 17. The morphogenetic tree. Arrows indicate causal links.

ordered cell differentiation (alias pattern formation), and in this specific instance with a 'tricolour' or French flag arrangement of three cell types.

We can re-draw the French flag model in more abstract ways – indeed this is desirable if the model is too complex for the task at hand. The same principle of maximum acceptable abstraction that we applied to organisms in Chapter 5 can be applied to models. If we wished to reveal the form the heterogeneity took, discrete versus continuous for instance, we would end up with the middle diagram of Figure 16. However, from a developmental viewpoint, what is of more interest is not the precise form of each heterogeneity but rather whether each is morphogenetic or not. It must be stressed that while only heterogeneities can provide morphogenetic information, not all of them do so. The adult organism, after all, is a great mass of heterogeneities, yet (with some exceptions) is morphogenetically inert. In the French flag model it is the flag itself that represents the adult organism and is morphogenetically inert. If we want to represent the model, then, in terms of a causal interaction between two morphogenetic heterogeneities and one 'terminal' one, the abstraction on the right of Figure 16 will suffice.

The French flag model is now clearly revealed for what it is: a *linear-sequence* causal model of development. In contrast to this, the morphogenetic tree, when expressed in the same way (see Figure 17), shows a *hierarchical* pattern of causality.

In fairness to both the proposer and the various users of the

French flag model, I should make it clear that it was not put forward to be an overall theory of pattern formation. Rather, if I understand it correctly, it was initially devised simply to illustrate how, in general terms, a gradient of diffusible morphogen could produce pattern formation in a two-dimensional array of cells. I have used it as if it were a general model of pattern formation in order (i) to make the point that the major theoretical problem in this field, as I see it, lies in determining the nature of the pattern of interconnection of causal links in development, and (ii) to argue that a linear-sequence pattern is too simple.

In contrast to the proposer of the French flag model, I *am* proposing that my morphogenetic tree represents a general theory of pattern formation, to be discussed in the next section. But before making such a bold claim in detail, it may be prudent to draw attention to some of the deficiencies in the very simple picture put forward in Figure 17. These are as follows:

1. There is not just a single trunk at the base of the tree; rather there is a multiple one. This seems inevitable, because it is inconceivable that only a single morphogen is present in the cytoplasm of the fertilized egg.
2. Bifurcation is not the only type of branching. More complex types (trifurcation and so on) may also occur. This will make the tree much more irregular.
3. Some branches die out before reaching the 'top', that is the adult organism. In other words, not all developmental processes going on at an early stage necessarily contribute to later ones.
4. There are no definite 'stages' of development, each with its characteristic number of heterogeneities; rather, the whole process is more akin to a continuum.

The above four complications are taken into account in the more realistic picture of the morphogenetic tree shown in Figure 18, which is beginning to resemble an *evolutionary* tree in some respects. As will be explained later, this resemblance may not be merely coincidental.

5. Even this more complex picture of the morphogenetic tree omits information on coordination and repeatability. A hier-

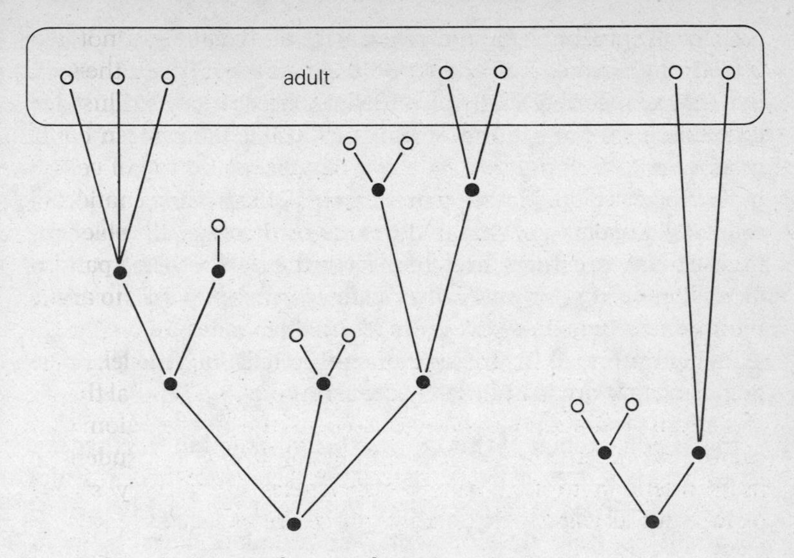

Figure 18. A more complex morphogenetic tree.

archical causal structure could certainly generate patterns, but whether it could do so in such a way as to ensure that, for example, the right arm is a mirror-image of the left arm (coordination) or that the whole outcome of the process is reasonably similar in different organisms (repeatability) is not so certain. I have the feeling that there may be some sort of cross-linking superimposed on the basic hierarchical structure, which would have a 'stabilizing' effect. This is not represented in Figure 18, basically because I have difficulty envisaging what form it would take.

6. The morphogenetic tree as so far stated is applicable only to organisms with simple life-cycles. (That is why I used the examples of snails and men when opening this section.) A system of 'nested trees' is necessary if we are to represent adequately the complex life-cycles found, for example, in insects. This idea will be expanded upon in the following section, as will point 7 below.

7. Insufficient attention has been paid to the causal links *between* generations as opposed to within them.

8. There are some differences in the form that development takes in plants, when compared to animals. I don't know whether these

are sufficient to make the morphogenetic tree idea inapplicable to them (which would be rather ironic, given the choice of name), but I think not. So I shall solve this problem by asserting that it *does* apply to plants, and waiting to see if botanists tell me otherwise!

9. I have chosen to picture pattern formation as being completed suddenly at some particular developmental stage. If, as seems likely, some structures are fully formed before others, pattern formation may 'peter out' rather than end abruptly. In that case, the tree's outline may look more like a diamond than a triangle standing on one of its angles. Something tells me that this particular complexity is of little consequence.

In the final section of this chapter I shall examine whether the morphogenetic tree concept represents a general theory of development. First, though, we need to look at the final two complications of how developmental information flows between generations, and between the different stages of a complex life-cycle, such as that found in insects.

Complex life-cycles and the generation bridge

If a creature such as a mammal has a morphogenetic tree like that shown in Figure 18 underlying its development, there is no way that the adult can initiate the developmental process in its offspring, as all the 'heterogeneities' that comprise the adult are terminal ones, that is, they are morphogenetically inert. Now in a sense this may be true of the male; but it cannot in any sense be true of the female, for, as we have seen, it is agents in the female parent's egg cytoplasm that initiate development in the progeny. What we need to do to take account of this recognized phenomenon is to represent the germ line separately, at least in the female.

A picture of the morphogenetic tree, modified in this way, is shown in Figure 19. Clearly, the morphogenetically active heterogeneities in the germ line cross the gap between consecutive generations, and act as the first important heterogeneities in the early development of the offspring. The number of these must of

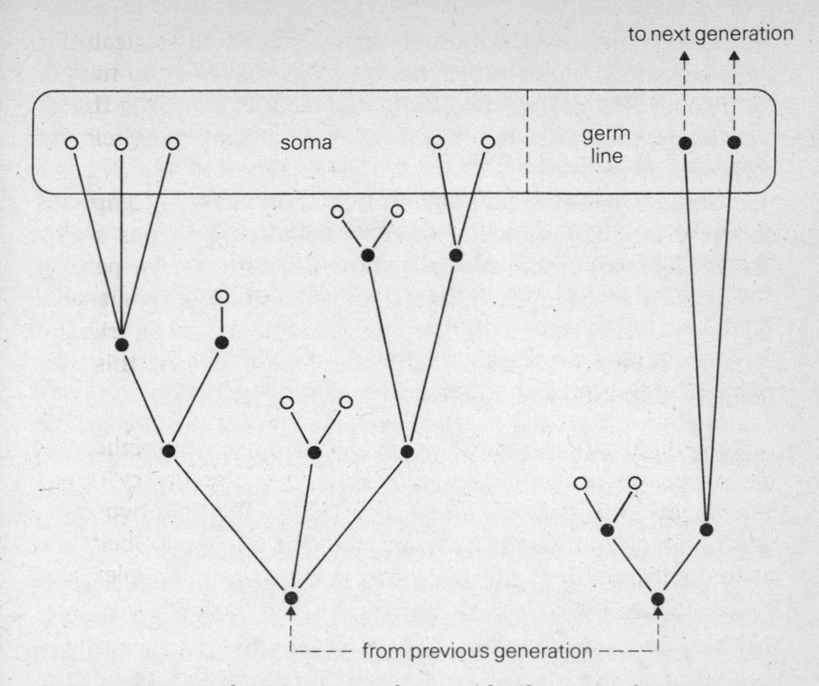

Figure 19. A morphogenetic tree showing the distinction between germ line and soma.

course be the same as the number of heterogeneities in the 'trunk' of the parent's morphogenetic tree, if the system is to remain constant from generation to generation, as it must. There is something reminiscent of Mendelian inheritance here – a sort of cytoplasmic counterpart, a 'particulate morphogen' concept if you like. The main difference, which is fairly obvious, is that the male contributes no 'particles', and there is thus no pairing comparable to what happens with genes.

The final problem for us to consider here is that of complex life-cycles. The general solution to the problem of how these can be visualized in terms of morphogenetic trees is, I think, to consider the developmental processes occurring in different life stages as separate trees which are linked together. A more detailed solution must be restricted to any one case – there is far too much variation among complex life-cycles for any detailed picture to be accurate for all of them. Examples of complex life-cycles include the egg/larva/pupa/adult system of insects, the sporophyte/gametophyte system of some plants such as ferns, and the multi-stage life-cycles of animal parasites such as the liver-fluke and its relatives.

I shall concentrate here on an insect example with which I am fairly familiar, namely *Drosophila*. This should be reasonably representative of the insects as a whole, or at least of the 'holometabolous' insects, where there is a rather drastic developmental transition at the pupal stage.

The question we need to pose is: how many separate developmental processes (and hence morphogenetic trees) are required to explain the overall development of a *Drosophila*? The possible answers range from one (a single egg-to-adult system) to five. Since there are three separate larval stages ('instars'), there are six life stages in the cycle (e, l_1, l_2, l_3, p, a), and hence a maximum of five inter-stage developmental processes.

What we know about *Drosophila* development, from a great variety of studies, suggests that the answer to our question is neither one nor five but two. The egg-to-larva transition, including all three larval instars, can be regarded as an integral developmental process. The building of the adult is also an integral process, but one which is quite distinct from the egg-to-larva system. Anyone who believes this last sentence because of a casual acquaintance with the insect life-cycle may be slightly surprised at the nature of the 'distinctness', for it is not merely distinctness in time. That is, the two systems are not separate simply because one precedes the other; and in fact there is considerable temporal overlap. Though it is not widely known out of embryological circles, the adult *Drosophila* is being built long before the pupal stage. As early as the first-instar larva, the adult exists as a group of 'imaginal discs' (small pieces of tissue destined to become the 'imago' or adult) which grow throughout the larval period, and from which, in the pupa, the adult is finally formed.

This two-stage process can be seen as two 'nested' morphogenetic trees, as shown in Figure 20. Since two nested hierarchies like this do also make up an overall hierarchy, it would not be inaccurate to consider even such a complex life-cycle as

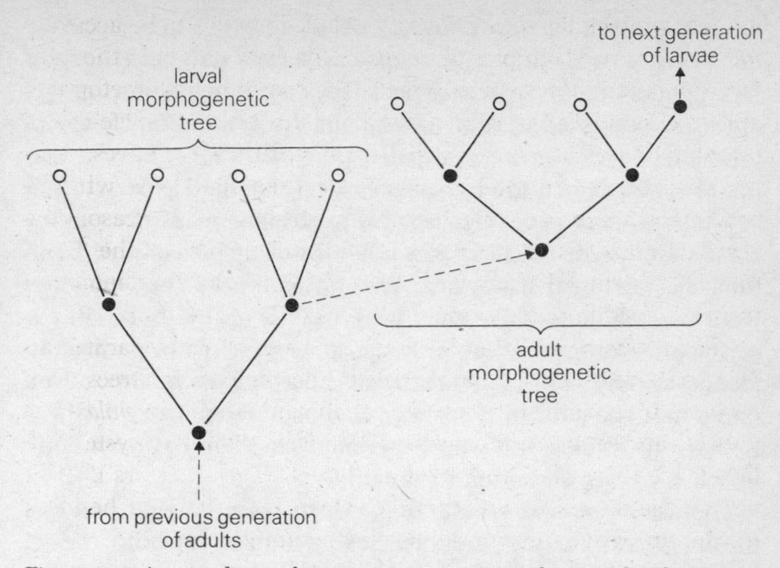

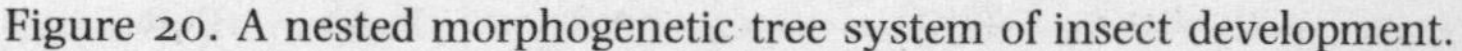

Figure 20. A nested morphogenetic tree system of insect development.

this as a single morphogenetic tree; but I think it is better to consider it as two nested sub-trees, as this emphasizes the high degree of independence of the two systems.

Is the morphogenetic tree a general theory of pattern formation?

We now turn from *description* of the morphogenetic tree to its *significance*. In particular, I want to examine the question of whether this idea constitutes a general theory of development, or at least of the major component of development that we call pattern formation. Strangely, it is not always clear, in science, what constitutes a general theory of some process or other, let alone whether or not it is correct. It is rare, at least in biology, for an idea of general significance to be immediately recognized as such, even though sometimes in retrospect it looks as if recognition was more-or-less instant.

The classic case of non-recognition of the significance of an idea is of course provided by Mendel, who in 1866 proposed the mechanism that we now know underlies almost all inheritance in the living world. Although there was thus a general theory of heredity during the last three decades of the nineteenth century, any biologist asked at that time about the particulate theory of inheritance would have responded with *'the what?'*. Now it was not that Mendel did not publish his ideas, for they appeared in a substantial (forty-page) article in an Austrian journal; nor can we claim that the use of a language other than English cut Mendel off from all intelligent biologists. And it was hardly that his ideas were too nebulous, since they were backed up by thorough experimental results. One can only speculate on why no one reading Mendel's paper appreciated its significance, but I would hazard a guess that the title of the paper had something to do with it. Roughly translated, this was 'Experiments in Plant Hybridization', which is hardly an ad-man's dream.

The use of an undramatic title, which essentially plays down the magnitude of the findings, has two quite separate effects. First, fewer people read the paper concerned. If, in scanning the contents page of a journal, I came across a paper with a title similar to Mendel's, I would continue scanning. If, on the other hand, I discovered a paper entitled 'A Revolutionary New General Theory of Inheritance', I would certainly read it. If we acknowledge that only a fraction of the people reading an important paper are capable of appreciating its significance, then the *total number* that read it may be crucial. If one in ten will understand the paper's importance and only two read it, the chances are that the advance buried within it will not be disentombed; while if two hundred people read it, the probability of the advance going unrecognized is negligible.

The second effect of the use of undramatic titles is to reduce the *fraction* of people who will appreciate a paper's significance by putting the onus of recognition on to the reader. That is, rather than the reader weighing up, and eventually making a decision on, a proposal for a particular general theory, he almost has to devise the theory himself from the experimental results (though he may be helped by claims within the text of the paper, which the author was not bold enough to embody in the title). Fewer readers will be capable of devising a theory, or of elevating its importance from an apparently minor claim half-way through the discussion, than will be capable of responding to a more boldly made claim, particularly a titular one.

Perhaps Mendel's work would not have gone unappreciated for thirty-four years if he had chosen a title analogous to Darwin's. He might, for example, have called his paper 'On the Mechanism of Inheritance by Means of Particulate Factors'. However, even Darwin's work was not immediately understood as having general importance, though the delay was much shorter, and is less well known. The first public airing of Darwin's views on evolution was a joint presentation (with a manuscript of Alfred Russel Wallace) read to the Linnean Society in 1858, and published in the Society's journal in the same year. Yet the Society's President subsequently reported that the year 1858 'has not been marked by any of those striking discoveries which revolutionize the department of science on which they bear' * – a wholly inappropriate conclusion! In this case, the recognition was delayed only for about a year. Darwin's *Origin of Species* was published in 1859, and was immediately appreciated as a book propounding a general theory of evolution, and reacted to as such.

Returning to the morphogenetic tree, and whether it constitutes a general theory of pattern formation, I have made sufficient claims for its importance so as, I hope, to prevent the burden of recognition from falling on the reader. However, there is another potential pitfall for those proposing new theories. That is, the theory may be seen as misguided or of little importance because it does not address the key issue. That raises the question of what the key issue of pattern formation actually is; which is not immediately clear. We should thus spend a little time addressing this question.

Three candidates readily suggest themselves for consideration as the key issue of pattern formation. These are:

1. The *initiation* of among-cell heterogeneity in the developing organism.

* For further details see *Neo-Darwinism*, by R. J. Berry (Arnold, London, 1982).

2. The mechanism by which developmental information is transferred from one heterogeneity to another.
3. The nature of the *pattern of interconnection* among the individual causal links (or heterogeneity transferences) that make up the complete developmental process.

We can immediately rule out the first of these, as we have seen that heterogeneity is present at all stages in the life-cycle, and can be re-created in an among-cell form after the egg stage simply by cell division. The second candidate, namely the mechanism of transference, cannot so easily be eliminated. However, I suspect, together with many other biologists, that no single mechanism (for example, a diffusible morphogen, as in the French flag) will be found to operate to the virtual exclusion of others – rather, a diversity of such mechanisms will probably be revealed when biophysical and biochemical studies on development have probed deep enough. So it may be that it will be impossible to generalize at that level. This leaves us with the final alternative, the pattern of interconnection among causal links, which is *independent* of the precise physico-chemical mechanism underlying any particular link. At this level, there is no reason why generalization should not be possible; and it is precisely this issue with which the morphogenetic tree deals.

I am certainly not the first person to propose that this 'pattern of interconnection' is the key issue of development, or indeed the first to propose that the answer is a hierarchical pattern, though the term 'morphogenetic tree' is mine. C. H. Waddington, one of the greatest thinkers in the field of developmental biology, stated that what we need 'is not so much a hypothesis of the ultimate physico-chemical mechanisms of development, but rather a *picture of the possible interactions* between developmental processes' (my italics). And as regards the form taken by this 'picture of interactions' or 'pattern of interconnections', several authors, including Waddington, have hinted at a hierarchical structure.

One such author was Arthur Koestler, a well-known broadly based thinker whose biological ideas were very heretical in relation to orthodox theory. For example, he referred to Darwinian

* This and a later quote come from *The Ghost in the Machine* (Hutchinson, 1967).

natural selection as one of 'the four pillars of unwisdom'.* Koestler was obsessed with hierarchies in general, and he saw development, like many other things, as a hierarchical process. For example, he states: 'Morphogenesis proceeds in an unmistakably hierarchic fashion.' However, we have to be rather cautious in interpreting this comment, and others like it, because it does not specify clearly in what *sense* development is a hierarchy. This is important, because in one sense development is very obviously hierarchical – it is a hierarchical pattern of cell descent, or a hierarchical cell lineage, if that is not a contradiction in terms. This fact in itself is obvious and uninteresting, in contrast to hierarchy of causal relationship, which is neither obvious (since the idea may be wrong) nor uninteresting (since if correct it comprises a general theory of pattern formation).

Because of the different meanings that can be attached to a statement that development is hierarchical, we need to look for statements that are more specific about the nature of the proposed hierarchy. One such statement can be found in a recent text by the British geneticist J. H. Sang. On the pattern of causal interconnections underlying development, Sang* says: 'Morphogenetic complexity depends on the activation/inactivation of gene sets, not single genes; *possibly in a hierarchical sequence of determinative steps*' (my italics). This is a much clearer, and thus more informative, proposition. It sounds very much as if Sang is proposing a morphogenetic tree system, though he makes little of it (the statement is buried in the middle of Chapter 13 of his book), and clearly regards it as a *possible* generality rather than an established one.

Having, I hope, persuaded the reader that the morphogenetic tree represents a general theory of pattern formation, the remaining question is whether or not it is correct. I leave this to others, as it is ultimately the scientific community at large, and not the proposer of a theory, that determines whether that theory is deemed to be correct. (This is not the same as *being* correct, though; there are plenty of instances of the scientific community becoming convinced of the validity of a new theory, only to find

* In *Genetics and Development* (Longman, 1984).

out later that it was wrong.) My only reservation about the morphogenetic tree is that it does not deal adequately with aspects of development other than pattern formation. I have particularly in mind here coordination and repeatability, as mentioned earlier, and also 'canalization', whereby development proceeds in a certain direction despite some genetic and/or environmental perturbations. Whether these phenomena can be pictured within a morphogenetic tree framework (by some sort of cross-linking for instance, as suggested in an earlier section) remains to be seen.

11 POPULATION REGULATION: AN ECOLOGICAL THEORY

We now make a conceptual leap into ecology: from developing organisms to populations of organisms. The reason why I have decided to cover an ecological theory at this stage in the book will not become clear until fairly far through the following chapter. For the moment, let's be content with the rationale that having spent the last two chapters discussing the problems of development in some detail, it will be refreshing to examine an apparently unrelated problem in a totally different area of biology. As John Cleese would say: 'And now for something completely different.'

Consider a hypothetical organism consisting of just a single cell and reproducing by binary fission once per hour. Starting with just a single individual, we shall have two after one hour, four after two hours, and so on. In general, after h hours we have 2^h individuals. So if our hypothetical organism is allowed to continue growing in this way, after ten hours there are 2^{10}, or more than 1000, individuals, and after twenty hours 2^{20} or more than a million. Elementary calculations of this sort greatly influenced Darwin, who provided some comparable figures for elephants. The basic type of growth just described, the biological significance of which was recognized by Malthus in the last century, is known as geometric growth. The term 'exponential growth', which is more widely used today, is strictly applicable to cases where populations grow continuously in time, rather than in discrete bursts corresponding to synchronized generations; but the basic idea underlying geometric and exponential growth is the same, and is indeed very simple. This is the idea that as time proceeds, not only does the population grow, but the rate of growth itself grows. This is shown graphically in Figure 21, from which it can

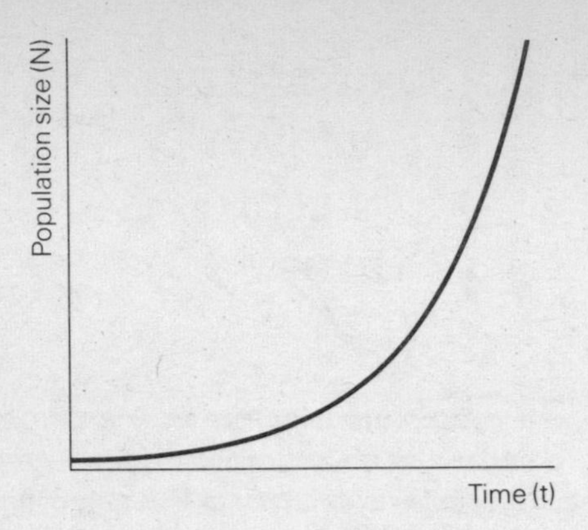

Figure 21. The exponential growth of a population.

be deduced that the population size rapidly tends towards infinity.

Exponential growth is always introduced in ecology textbooks as a sort of straw man – that is, something set up to be immediately burned down. Its importance lies not in the fact that populations grow exponentially, but rather in the fact that they don't. This leads us to the question of *why* they don't, despite having *potential* for growth of this kind, which is really the central issue.

You can approach the same problem in a different way. Suppose you are an ecologist who has been estimating the population size of a particular species in a particular place. This could be water snails in a lake, owls in a forest or buffalo on a large plain. Studies of this sort, conducted over many years, usually reveal that the population concerned fluctuates within limits that are narrow in relation to what is theoretically possible. For example, a population might well fluctuate over an order of magnitude, say between 1000 and 10,000, rather than right through the range one to infinity. The central question then becomes: what constrains the population size to the relatively narrow band of values within which we usually find it?

It will be useful to consider first what happens in the comparatively simple situation of a population growing in a container of some kind in the laboratory. Population experiments of this sort are often done with tubes of micro-organisms, jars of beetles,

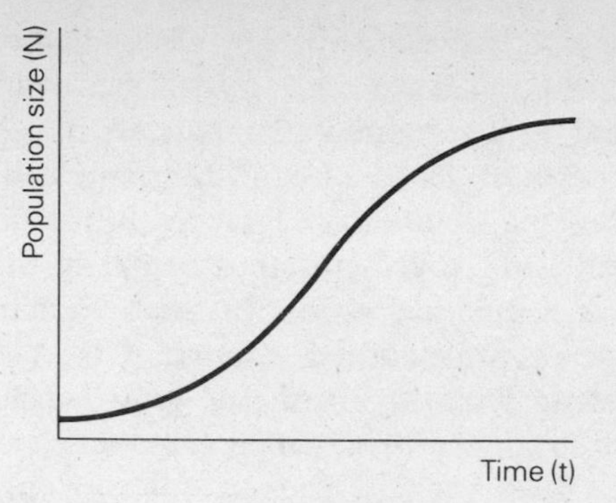

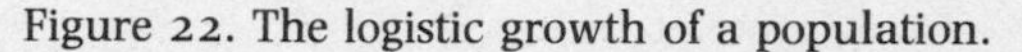

Figure 22. The logistic growth of a population.

cages of flies and so on. The species are usually chosen simply because of ease of handling and short generation time.

If we look at the pattern of population growth in one of these experiments – say in the case of the unicellular organism *Paramecium*, worked on by the Russian ecologist G. F. Gause – we find that it approximates to the pattern shown in Figure 22. This S-shaped or sigmoid pattern, when applied to the growth of biological populations, is often referred to as logistic growth.

The important point about logistic growth is that, while it starts out accelerating, like its exponential counterpart, it ends up decelerating and eventually becoming static – that is, the population eventually settles down to a fixed size. This size is often referred to as the *carrying capacity* of the culture vessel for the species concerned – that is, the maximum number of individuals of that species which it can sustain. For technical reasons, higher organisms, such as insects, do not conform as closely to the logistic pattern as simple ones like *Paramecium*, but they still approximate much more closely to this pattern than to the exponential one.

In simple laboratory systems of this kind, it is fairly obvious why population growth slows up and eventually stops. Whatever container is used has a certain limited amount of food which is periodically renewed by the experimenter; and this limited amount of food determines the carrying capacity. If the actual population size at any one time is beneath the carrying capacity, the population will grow; otherwise it will not. Occasionally, with certain designs of cage (or bottle or tube), physical space rather than food will be the limiting factor; and with plants, we have to distinguish between the different 'foods' of light, nutrients and water. However, all of these possibilities are included in the term 'resource-limited growth', where growth stops owing to the using-up of something the population needs. In the laboratory, growth is nearly always resource-limited, though it is conceivable that build-up of waste products could stop growth before the population has got to the size where food is limiting.

From laboratory to nature

Can we apply this conclusion of near-universal resource-limited growth to populations living in natural habitats? Well, the answer is 'maybe', but certainly not straight away. We cannot immediately proceed from laboratory conclusion to general conclusion here because laboratory systems are deliberately simplified in ways which may affect how population sizes are limited. For example, in the laboratory, we exclude predators and, if possible, parasites of the species in which we are interested. Further, physical variables such as temperature and humidity are usually held constant as far as our equipment (incubators, for instance) will allow. Because of these procedures, neither predators nor drastic changes in temperature are permitted to affect the size of our population. In nature, not only may both these things have an effect, but the two main alternative theories to resource-limited growth are based on (a) predators and parasites, and (b) fluctuations in climatic variables such as temperature.

So far I have talked as if there were two types of limitation to which a population may be exposed – 'resource' and 'other' – the latter including the theories of predator-limitation and climate-limitation, as I shall call them. This is a misleading grouping of the three theories, because in fact resource-limitation and predator-limitation have something in common, while the climate-based theory is fundamentally different. That is, instead of an interrelationship structure of

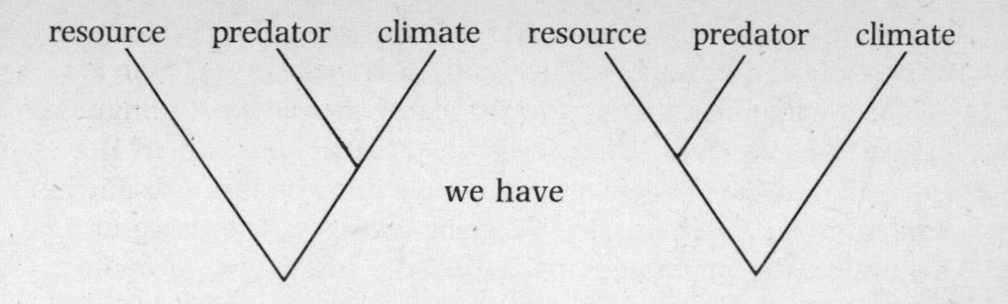

So what is it that the resource- and predator-limitation theories have in common? The answer is that they are both based on the idea of density dependence, which means that we need to see what is meant by this phrase before we can proceed any further.

Interpreted literally, this phrase implies that something depends on, or varies with, density. That something is the per capita rate of population growth. But the phrase should not in fact be interpreted literally, because it means something more restricted than simply that the per capita growth rate varies with density; it must vary *negatively* with density. In other words, when we say that a population exhibits density dependence (or density-dependent growth), we are saying that as that population becomes more dense, the per capita growth rate declines.

Actually, logistic growth as shown in Figure 22 necessarily implies density dependence. However, the logistic curve is neither a necessary nor even a very helpful picture to use for the illustration of this phenomenon. A better picture is that of Figure 23, where the two variables of interest are plotted directly against each other. Any *descending* line or curve, such as the one shown, represents density dependence. What we see, then, is that at low density the population grows rapidly. As density increases, the growth rate (a) becomes slower, (b) reaches zero and (c) goes negative, that is we reach a stage where the population begins to decline.

It is worth spending a while looking at Figure 23 and making sure you understand it; experience suggests that it is open to frequent misinterpretation. One important point is that it shouldn't be considered as a time series – a population starting off small, growing, getting bigger, growing more slowly and so on. In this case, the population size would become static once a

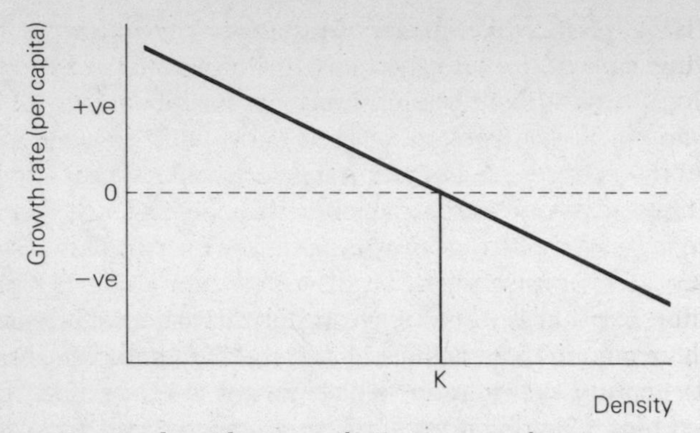

Figure 23. Density-dependent growth. K represents the carrying capacity. The area occupied (A) is assumed to be fixed at 1 unit, so that density (N/A) and population size (N) are numerically equal.

certain size was reached (as it does in logistic growth); consequently, 'negative growth' would not be possible. But if you imagine instead the experimental manipulation of density to a variety of levels, including some in excess of the carrying capacity, followed by observation of the direction and magnitude of growth subsequent to the manipulation, then it is indeed possible to build up a more complete picture of the relationship between density and growth rate, as seen in Figure 23.

The most important thing about density dependence, as illustrated in the figure, is that it gives rise to a stable equilibrium in population size – that is, a certain fixed number of individuals to which any other population size will converge. Larger populations will decline and sink down to the equilibrium number, while smaller ones will grow up to the equilibrium. This kind of behaviour of a system is of much general importance both in biology and in other fields. It is variously referred to as homeostatic, equilibrium-seeking or negative-feedback behaviour; and we met a form of it before in the bacterial operon system (p. 114).

Now as regards the *cause* of density-dependent population growth, the most obvious one is resource-limitation. In populations inhabiting an environment, whether in laboratory or field, where food is limiting, small populations grow simply because food is plentiful, while large ones (temporarily in excess of the carrying capacity, owing perhaps to immigration from neighbouring populations) decline simply because food is insufficient to maintain the current number of individuals. However, food-limitation (or resource-limitation in the broader sense) is not the only biological source from which density dependence can arise.

The main alternative source is predators (or parasites). Subject as they are to natural selection (as is everything else in biology), predators tend to feed in ways which are fairly efficient, and which result in their acquiring more prey (and hence energy input) per unit effort (and hence energy output) than if they hunted for their prey at random. Thus a predator will not waste much time hunting in a sparse prey population if it can turn either to a denser population of the same prey species in a different locality or to a dense population of some other prey species in the same locality.

If a population thus suffers negligible predation when its numbers are low, but disproportionately heavy predation when its numbers are high, this is obviously a form of density dependence. Whatever the growth rate of the sparse population suffering little predation, it will clearly decline as the population suffers increasingly severe predation as it gets denser. It may be that a density is reached where deaths from predation cannot be compensated for by births, and in this case the population will decline. Thus we have the same equilibrium-seeking behaviour as can arise from food-limitation.

One additional point worth making here is that a predator-mediated equilibrium is necessarily lower than a food-mediated one. A predator cannot make a population reach an equilibrium which is in excess of its carrying capacity; but it certainly can prevent it from ever reaching that capacity, if the 'negative growth' phase resulting from severe predation occurs at intermediate densities.

A population tending towards an equilibrium owing to density dependence (whether resulting from resource- or predator-limitation) is said to be *regulated*. Alternatively, the phenomenon itself is referred to as population regulation. Two points should be noted here: (1) Regulation is an entirely natural phenomenon, distinct from any attempts by man to 'regulate' population numbers (for example of crop pests) to certain (usually low) values. Because of potential confusion here, such attempts by man are labelled 'control' rather than regulation. (2) The climatic theory of population growth (and decline) in natural populations does not involve regulation, nor the density dependence underlying it.

What is this third theory? Well, put simply, it states that in nature populations never grow for long enough to reach the limits of their food supply, or even to reach a level at which predation could cause them to decline. Rather, the view taken is that, after perhaps a few generations of growth, there will be a drought, a severe winter or some other climatic problem which will cause numbers to decline. When the crisis is over, the population will resume growing again, from a low starting point, and will continue to grow until the climate once again becomes unfavourable. The population size therefore lurches about in a haphazard way, largely resulting from the action of agents of mortality (such as droughts) whose effect is *not* dependent on density.

Distinguishing between the theories

As usual in science, we like things to be as simple as possible. So it would be nice if all natural populations, of whatever species, and in whatever place, behaved in accordance with one of our three theories. Of course, we know enough by now not to expect that 100 per cent of populations will do this, but it might not be too optimistic to suspect one of the theories to hold for 90 per cent or more of populations, and the debates that have been (and are being) conducted on this issue have been dominated by disputes among the opposing camps of optimists, each thinking that it is *their* theory that applies to the vast majority of populations.

Unfortunately, it is *not* possible to distinguish between the three theories simply on the basis of monitoring population sizes over several generations, and looking at the pattern of growth (or decline). Figure 24 shows three hypothetical populations, each behaving in accordance with one of our three theories. Ecologists

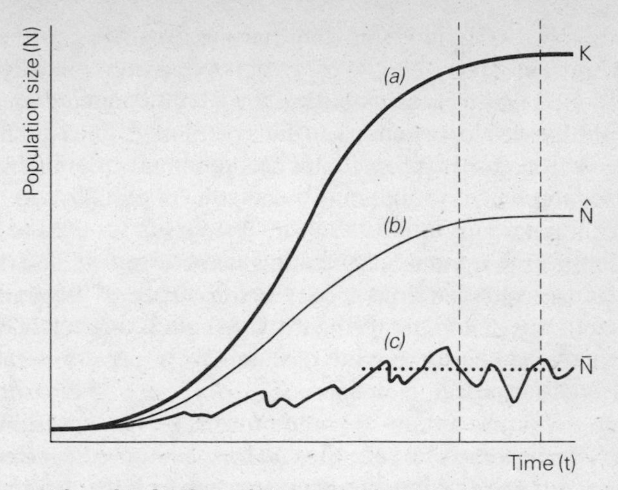

Figure 24. Three kinds of population growth. (*a*) Growth up to equilibrium (K) with limiting food supply. (*b*) Growth to lower equilibrium (N̂) set by predator. (*c*) Growth under fluctuating climatic conditions, giving permanently sub-equilibrial population size (N̄).

rarely observe all stages of population growth from the original establishment of the population, so what we have to interpret is usually a restricted set of data such as that between the two vertical lines in the figure.

The two things that look as if they might be usable in distinguishing our three theories are (a) the average population size over the period of observation (since food-limited is higher than predator-limited, which is higher than 'climatic'), and (b) the amount of fluctuation around that average, since the climate theory explicitly invokes such fluctuation, while the others do not. In fact, neither of these apparent distinctions is useful. The position of the average population size is uninformative, because in any one population it simply takes a particular value, and we have nothing with which to compare this value. There is no *a priori* reason to expect any of the three theories to produce any particular average population size – all we know is that these averages have a necessary *ranking*, which is of little value when we have only one average, and hence nothing to rank.

As regards fluctuations around the average, the figure shows these to be absent in the two types of population regulation, and present in the case of the climatically governed population. However, the figure was intended to illustrate only the core of each theory. In practice, no proponent of population regulation believes that populations in nature reach a nice neat equilibrium value. Rather, fluctuations around the equilibrium are thought to occur for all sorts of reasons – including climatic ones. Theories of population regulation do not deny the existence of these fluctuations, but they tend to disregard them and to deal with the underlying equilibrium, which is due to a density-dependent factor – either predators or food.

Now although certain apparent means of distinguishing between our theories cannot in the end be used, there are other, less obvious ways of making the necessary distinctions. One kind of study that has been productive is the search for density dependence in natural populations. If it can be demonstrated that the percentage mortality increases with density in these populations, or that the number of offspring per individual decreases (or both), then we have demonstrated the density dependence that must be present for there to be an equilibrium. Most studies which have involved a search for density dependence have been successful in finding it, and as a result most population ecologists are now agreed that the vast majority of natural populations are regulated – though they are also exposed to additional fluctuations of a non-regulatory nature. Admittedly, a few ecologists still cling to a predominantly climatic, non-equilibrial view of populations, including the Australian ecologists Andrewartha and Birch who were instrumental in developing this view in the first place. But I think that this camp can now be regarded as so small a minority that we can consider one aspect of the problem – namely whether populations are regulated or not – as settled in favour of the former view.

This still leaves the question of *how* populations are regulated, that is whether, within the regulation theory, the predominant mechanism is regulation from 'above' (by predators) or from 'below' (by food). There is currently much debate on this issue. My own view is that in this instance the diversity of the natural world will preclude a simple answer – that is, I don't think that all (or even 90 per cent) of populations will turn out to be regulated in one of these ways. I may be wrong here, and indeed, as one who appreciates simplicity, I would in some ways be pleased to be wrong – but this remains to be seen. At least this continuing controversy is *within* the regulatory school, so whichever way it turns out we shall still have a general theory of population regulation.

From one species to two

Suppose that, inhabiting a particular area, we have a resident population of 'species A', which is limited in size by the availability of a particular food resource. Now suppose that a small number of individuals of a closely related species ('species B') migrate into the area, thereby founding a small population. If this new population of species B depends on, and will ultimately be limited by, the same resource that limits species A's population, what will happen now that we have two 'jointly limited' populations?

Anyone without an ecological background could be forgiven for thinking that both species would persist in the area, but with the numbers of each being lower than they would have been had that species been present on its own. However, this does not happen. Instead we get a process called *competitive exclusion*, wherein one of the species is gradually eliminated from the area concerned. The 'winning' species is whichever is better adapted to the prevailing conditions, and this will in some cases be the resident, in some cases the immigrant.

The process of competitive exclusion can readily be demonstrated in two-species laboratory systems. Indeed, experiments in which two species compete for a common limiting resource, with the outcome that one of them eliminates the other, are among the most repeatable of all ecological experiments. Nor is this phenomenon a laboratory artefact: the simplest mathematical model that can be devised to describe competition between two species predicts just such an outcome; and various kinds of evidence from natural populations point to the previous

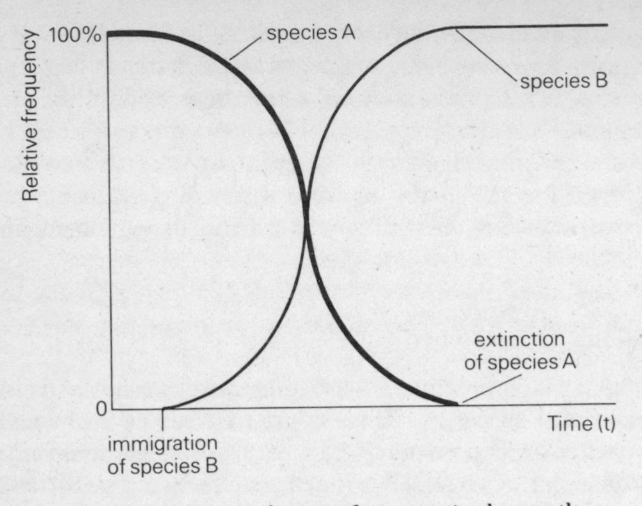

Figure 25. The competitive exclusion of one species by another.

occurrence of competitive exclusion as one determinant of present-day species distributions – though as usual the data on *natural* populations are the most difficult to interpret.

Let us suppose that in the case of resident species A versus immigrant species B with which we started, species B turns out to be superior. In this case, the pattern of replacement of species A by species B will be as shown in Figure 25, where the frequency of the new, superior species is seen increasing from an initially low value to, eventually, 100 per cent.

We sometimes speak of the *competitive ability* of one species relative to another in this general context. In our specific (and hypothetical) example, it is clear that species B has a greater competitive ability than species A. Now this ecological terminology is appropriate since we are dealing exclusively with an ecological situation. Or are we? Recall the case of industrial melanism in *Biston betularia* discussed in Chapter 6. What happened there (in urban areas) was that a new type, superior in the prevailing conditions, entered a population of an 'old' type at low frequency, then spread gradually through the population, eventually (in some cases) approaching fixation. If we chose to call the new type or melanic 'type B' and the old type or non-melanic 'type A', then Figure 25 applies just as well to this situation as it does to the one with which we started.

In other words, looked at from a certain viewpoint, the process of elimination of one type of organism by another through natural selection acting *within* a species, and the process of exclusion of one type of organism by another through competition acting *between* species, are one and the same thing. Of course, from other viewpoints they are not; the most obvious difference being the interbreeding of the two types in the first case but not in the second – but this does not affect the nature of the replacement that takes place.

In both situations – competition between species or between genetic variants within a species – the weaker type of organism can persist if it has some means of escaping from direct competition with the superior type, such as having a slightly different ecological niche. For example, the two types may be limited by two subtly different kinds of resource rather than the same resource. Again, we have a parallel between the 'genetic' and the 'ecological' situation.

What all this tells us is that it is joint limitation of population size rather than interbreeding that dictates when we should measure the fitness (or competitive ability) of one type of organism relative to another. This applies regardless of whether limitation is by resources (as in my example above) or by predators. If two types are regulated by different factors, each will be able to exist regardless of the presence of the other. If they are regulated by the same factor, only one will survive. This point has implications for evolutionary theory, which has traditionally regarded interbreeding as the criterion for determining fitness in a relative rather than absolute way, and has regarded only relative fitness as being important. These implications are discussed in the following chapter.

12 EVOLUTION OF SPECIES AND HIGHER TAXA

There is much current debate on whether the 'micro-evolutionary' studies of population geneticists, which deal with minor evolutionary changes occurring *within* present-day species, provide the whole story (or even an important part of the story) of 'macro-evolution'. This latter term is used to refer to long-term evolution, involving as it does the origin of new species as well as groupings higher than species in the taxonomic hierarchy. The importance of Darwin in evolutionary theory is tightly linked to the resolution of this debate, because Darwin strongly advocated the view that macro-evolutionary changes, including the 'origin of species', were brought about merely by the accumulation of many smaller changes occurring over shorter periods.

This debate is as old as the theory of natural selection itself. Darwin's rejection of the idea that large evolutionary changes could sometimes occur all at once, as a sort of 'evolutionary jump' rather than via an accumulation of many smaller changes, was questioned by his otherwise supporter T. H. Huxley (who gained the nickname of 'Darwin's bulldog' because of his strong defence of the idea of natural selection). In the 1940s the geneticist Richard Goldschmidt claimed that macro-evolution involved major changes that were fundamentally different from those of micro-evolution. Recently, some (but not all) of the supporters of the theory of punctuated equilibrium have taken a broadly similar view, though no doubt they would largely reject the actual mechanisms of macro-evolutionary change proposed by Goldschmidt, which now seem untenable to most biologists.

There is a danger of starting the discussion of this whole issue off on the wrong foot. The danger lies in recognizing only two levels of evolutionary change – micro and macro. Whether or not the origin of a new species requires an explanation additional to, or different from, the kind of natural selection that operates within species is quite a separate issue from that of whether the origin of a major group (at the class or phylum level in taxonomy) such as the vertebrates, the gastropods (snails) or the insects requires such an explanation. Thus it will be helpful if we restrict 'macro-evolution' to cover the origin of species, and use 'mega-evolution' to refer to the divergence of the major taxa. This is not a new proposal – in fact it was made in 1944 by the American palaeontologist G. G. Simpson.*

If we adopt a three-way split between micro-, macro- and mega-evolution, theories on evolutionary mechanisms may be distinguished by where (if anywhere) they propose a mechanistic discontinuity. Although four kinds of theory are imaginable, in practice we get only three. The first is represented by Darwinian theory, which proposes no discontinuity and claims that the mechanism of evolution is the same at all three levels. The second type of theory is that which lumps macro and mega together, and sees these as differing in mechanism from micro. Goldschmidt's theory, and some versions of the theory of punctuated equilibrium, fall into this category, as do the now-discredited views of Hugo de Vries. The third type of theory puts the discontinuity between macro and mega, thereby lumping macro and micro together. Simpson himself thought along these lines, and what follows in this chapter will show that I have some sympathies with this approach. These three types of theory may be represented, then, as:

'Darwinian'	micro–macro–mega
'Goldschmidtian'	micro | macro–mega
'Simpsonian'	micro–macro | mega

where the vertical bars represent discontinuities and the horizontal dashes indicate a continuum. The labels for the three theories are a bit arbitrary, but they will suffice for the moment.

Because I see them as separate issues, I shall treat the origin of species and the origin of higher taxa separately; and the rest of this chapter will thus be divided into two 'sub-chapters', A and B, dealing with 'macro' and 'mega' aspects respectively. In A the focus will be largely palaeontological, and I shall concentrate on

* In *Tempo and Mode in Evolution* (Columbia University Press, New York).

the recent theory of punctuated equilibrium. In B I shall take a more developmental approach. As regards the evolutionary time-scales involved, species tend to persist for about 1 to 10 million years (MY), while classes and phyla originating 500 MY ago are still with us, so there is clearly a large difference in 'temporal focus' between parts A and B.

A. MACRO-EVOLUTION: THE ORIGIN OF SPECIES

Recent detailed studies of the fossil record, in which the evolutionary fate of a particular species is concentrated upon, have revealed that a common pattern of change is a 'rectangular' one, as illustrated in Figure 26, which shows the origin of a new species occurring about half-way through a 2MY fossil sequence. Prior to such detailed palaeontological studies we had no clear picture of what 'shape' evolutionary changes over these long

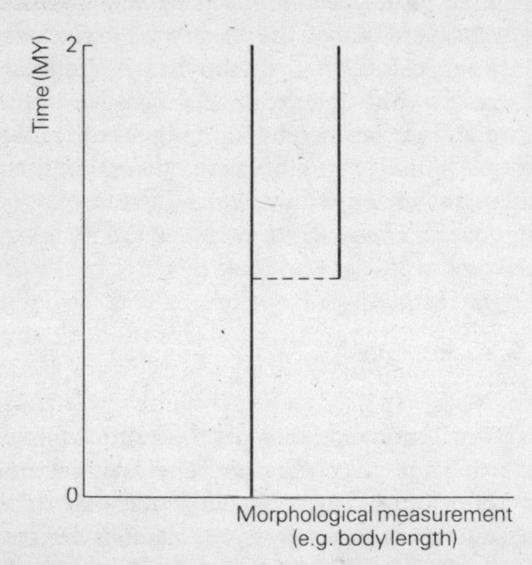

Figure 26. The origin of a species, as observed in a sequence of fossils. The solid lines indicate the value of the measured character. The dashed line indicates the presumed evolutionary relationship. The pattern is 'rectangular' (see text) only if this link is included; otherwise it is just a pair of parallel lines with different starting points.

periods of time would take, but it was widely believed that a much more gradual change would be the norm.

'Rectangular' patterns of change are now known to be common, though not all changes are of this kind. But it is not so clear what interpretation should be made of these patterns – that is, what mechanism underlies them – and whether, as some have claimed, they are evidence of some sort of non-Darwinian evolution. (Darwin's own diagram of evolutionary change in *The Origin of Species* was much more gradual and tree-like.)

Part of the problem in interpreting the rectangular or 'punctuational' patterns lies in the fact that the supporters of the 'theory' (whatever it is) themselves make different interpretations. Perhaps the best way to approach the problem, then, is to think of all possible mechanisms that could underlie the punctuational patterns, and to evaluate each of these for ourselves.

The first possibility is that what we are observing is the 'creation' of a species rather than its evolutionary origin. Now this may seem ridiculous to many readers, yet the creationist lobby is quick to latch on to any apparent difficulty in evolutionary theory, particularly one in which species arise suddenly, and they have not failed to interpret punctuations in the fossil record in this way. However, I shall dismiss this viewpoint for the moment. Creationists are not scientists (whatever they may say) and their ideas are not appropriately discussed in a 'scientific' chapter such as this. Their turn will come in Chapter 14.

The second possibility is that species really do remain constant in bodily structure throughout their duration, changing only during episodes of splitting of lineages (that is, episodes of speciation or cladogenesis). This view is certainly erroneous. What is observed in a fossil sequence *from a particular place* is the evolutionary history of a population, not the whole species to which that population belongs. We know very well that living species often exhibit recognizable sub-species or races in different parts of their range. So, at the very least, a species undergoes morphological differentiation as it spreads from its point of origin to occupy a wider geographical area.

Recognition of the spatial aspect of the problem should help us to formulate a more reasonable view of punctuated equilibrium.

Perhaps we could say that the typical species undergoes marked morphological change only during speciation and the phase immediately after this, when it spreads to occupy a wider range. Populations rapidly adapt to each local area, and having done so remain fairly constant in morphology thereafter – until the next speciation event. I imagine that this view would not be found too objectionable either by the proposers of the theory of punctuated equilibrium – the American palaeontologists Niles Eldredge and Stephen Gould – or by population geneticists, to whom many other versions of punctuated equilibrium are unacceptable. (Like any view, though, it is bound to be objectionable to someone!)

Although one can perhaps engineer a measure of agreement on the broad interpretation of the pattern by stating it in this way, there is still the question of what *mechanisms* underlie it. Of course, the mechanisms involved in causing punctuations must be different from those involved in maintaining the long periods of relative stasis between consecutive punctuations – and indeed there has been some discussion of which of these is more important. But clearly both are of interest, as they both contribute to the overall pattern.

Patterns of punctuated equilibrium in the fossil record can in fact be interpreted very easily within the conventional Darwinian framework, as elaborated by population geneticists. One common view of how species originate is that they do so by 'allopatric speciation'. 'Allopatric' simply means 'different areas' and the idea is that a population inhabiting an isolated, far-flung corner of a species' geographical range, where local ecological conditions are unique, evolves in a different direction from the rest of its species in order to adapt to those conditions and, as a spin-off of this divergent evolution, becomes reproductively isolated from the other members of what used to be the same species, and thus becomes a new species in its own right. At some later stage, our new species may re-migrate into the area occupied by its parent species, to coexist with it, to exclude it competitively, or to be competitively excluded by it, depending on their precise ecological relationship.

Suppose the new species competitively excludes the old in a particular area. A core taken for analysis of fossils in this area will reveal a rectangular pattern, but the cause of the punctuation is an ecological one (migration and competitive exclusion), not an evolutionary one. If the area which our speciating population inhabited at the time it actually underwent speciation was small relative to the total geographical range, *most* areas would show these 'ecological punctuations'.

What if we *knew* that our core of geological material in which a punctuation occurred had been extracted from the area of speciation itself? Would this then be inexplicable in terms of standard Darwinian selection? In fact, it would not, and the apparent problem of producing a significant morphological change 'instantaneously' is solved when we realize that what is instantaneous geologically may be extremely prolonged ecologically. Recall that the time-scale in Figure 26 extended over a 2MY period. The punctuation might have taken two years, but it might also have taken 2000, or perhaps even 20,000 – still only 1 per cent of the total. We cannot usually resolve these events any more precisely than this. Now while 20,000 years is a tiny fraction of a two-million-year period, it is a vast amount of ecological time in which for natural selection to act; and there is no problem whatever in producing the kind of changes observed, by the standard Darwinian selective process, in this sort of period.

So much for punctuations, but what about stasis? Again, it is easy to envisage a standard selective solution to this problem. If a population is well adapted to its local environment, selection still acts upon it. But in this case, rather than acting in favour of one extreme of the distribution of a character (such as size) and against the other, it acts in favour of the middle of the distribution and against *both* extremes. The result of this kind of selection (which is called normalizing or stabilizing selection) is morphological constancy rather than morphological change.

I have developed a 'conventional' interpretation of patterns of punctuated equilibrium, in which these patterns are seen as interesting phenomena which expand our knowledge of the details of evolutionary change, without requiring the operation of any fundamentally new evolutionary *mechanism*. Is such a view acceptable to the group of palaeontologists who have championed, and supported, the 'rectangular' view of evolution?

I must admit to being unsure of the answer to this question, though I suspect that some who fall into this category would accept an explanation broadly along the lines described above, while others would not. Some of the punctuationalists talk more in terms of developmental change and developmental constraint than in terms of natural selection; but whether there is a conflict here depends on the precise meaning of these terms. If the structure of the adult organism changes rapidly (in evolutionary terms) over a period of a few thousand years, clearly this is also a period of intense developmental change, because the only way in which adult structure can be altered is through alteration of the developmental process from which it stems. Similarly, constancy of adult form over a long period of time implies constancy (and therefore constraint of some sort) of development. Thus the 'developmental' and 'selective' approaches *can* be one and the same thing. However, they need not necessarily be so. Consider the term 'developmental constraint'. This could imply one of two things. First, that there is constrained variation in development because it is somehow impossible for the developmental pathway to deviate from a narrow range of directions without making a dramatic switch of some sort; that is, development is not infinitely plastic, but there are 'gaps' between alternative possible developmental routes. Second, that the range of developmental variation is constrained by normalizing selection eliminating extreme variants – that is it was possible for the variants to be constructed, but they were less viable than intermediate types and were thus removed from the population by selection.

The second of these interpretations of 'developmental constraint' is clearly quite compatible with the conventional, selectionist view; but the alternative interpretation is not. If there are indeed developmental constraints of the 'internal' sort underlying long periods of stasis in the fossil record, then the patterns of punctuated equilibrium *are* telling us something new about evolution – and, for that matter, about development – albeit what this new 'thing' is is not yet clear.

I have to admit that I am uncertain whether the conventional selective view of punctuated equilibrium and of developmental constraint *at this level* is correct, but I am convinced that, in general, the view of development being infinitely plastic is wrong. However, I think that the most important 'gaps' between alternative developmental pathways are much broader than those we have been discussing (as are the viable routes that they separate), and that where we see these in evolution is in the gulfs between the major taxa (such as phyla) rather than the tiny crevices (if these even exist) between species.

This is a personal view, and it is clear that this issue remains to be resolved. However, before leaving the area of punctuated equilibrium, two things remain to be discussed. First, we have been dealing entirely with morphological change, which is appropriate, since it is this that we see in the fossil record. Yet evolutionary changes are necessarily *genetic*, and not all morphological changes have a genetic basis. Studies on living organisms show quite clearly that large morphological changes can be produced directly by changes in the environment. Now this is not Lamarckism – the changes are not inheritable – yet if the environment changes from one stable state to another, morphological characteristics, such as the shape of a shell, may do likewise, without being accompanied by any evolutionary change at all. It is difficult, if not impossible, to separate these *ecophenotypic* changes from genetic ones in palaeontological studies, and some of the former must masquerade as punctuations. I don't suppose for a moment that all punctuations are of this sort, but I'm equally sure that some of them are!

The final thing deserving mention here is the theory of species selection, which has been closely associated with that of punctuated equilibrium, though the two are not inextricably intertwined. Species selection, first proposed by the American palaeontologist Steven Stanley, is essentially a direct analogue of Darwin's 'individual selection' operating at a higher level. Stanley suggests that we view origination (through speciation) and extinction as being to the species what birth and death are to the individual. Just as Darwinian natural selection involves the proliferation of individuals that leave most progeny at the expense of others, so species selection involves the spread of groups of species (clades) within which speciation rates are high.

It is difficult to claim that this theory is wrong, and indeed, like natural selection itself, it comes close to being a tautology. Two restrictions to the theory deserve mention, though. First, it does not tell us *why* certain groups have higher speciation rates (or lower extinction rates) than others. Second, it cannot explain the origin of evolutionary novelty – only the proliferation of species already possessing some novel characteristic. The importance of the theory of species selection lies in its ability to bring the evolution of the groups immediately above species – genera and families – into our combined micro/macro-evolutionary picture.

B. MEGA-EVOLUTION: THE ORIGIN OF BODY PLANS

Despite the significant changes in morphology at the species level implied by the 'rectangular' pattern illustrated in Figure 26, these changes are slight in relation to the grand scheme of things. Indeed, some species – called sibling species – show hardly any differences in morphology at all, despite their lack of interbreeding. (These, of course, will *not* give a punctuational pattern in the fossil record – rather, their divergence will probably go undetected.) Even where there are interspecific differences that are readily discernible by a taxonomist, they are often slight enough to be undetectable to the untrained eye. Differences between genera or families, while greater, are still usually of a quantitative rather than a qualitative nature. The morphological differences between man and chimp, for example, are of this sort. We both contain the same tissues and organs, as well as the same overall structure; our differences are merely those of shape. (Obviously, in this particular case, the behavioural differences are enormous – but that is a separate matter.)

As we ascend the taxonomic hierarchy, we eventually reach a point where differences are (or appear to be – and here is an important argument) discrete or qualitative rather than merely quantitative, such as those of size and shape. The point at which we reach these discontinuities varies, but it is usually at the levels of order, class or phylum. Since these labels are rather uninformative to the general reader, a few examples will help. The body plan of the Lepidoptera (butterflies and moths) is quite distinct from other insect body plans such as that of the Diptera (flies). Houseflies, *Drosophila* and seaweed flies are all clearly recognizable as flies, and red admirals, cabbage whites and meadow browns are all very obviously lepidopterans. Each group (in this case they are orders) exhibits internal quantitative variation, but seems to be separated by a distinct 'gulf' from other groups at the same level. (Other insect orders include Coleoptera (beetles) and Hymenoptera (wasps and bees).)

An example of discrete differences at the level of the class is provided by the mollusca. Here, the major classes of Gastropoda (slugs and snails), Bivalvia (cockles, mussels, etc.) and Cephalopoda (octopuses and squids) are fundamentally different in their design. Finally, ascending to the phylum level, any of the major phyla seems to differ in a discrete manner from others. For example, it is known from comparative embryological studies that the closest relatives to our own phylum Chordata are the starfish and their allies – phylum Echinodermata. It is hard to imagine two more different designs, at least at the level of adult structure.

Discrete differences of these kinds separate what are known as body plans – the major structural themes in the living world. It is clear that the exact nature of a body plan is difficult to pin down, and that this idea does not always connect with the same level of taxonomic grouping. This has led some biologists to reject the idea altogether, but others, including myself, regard it as useful, perhaps even essential. In fact, the idea of a body plan is implicit at all levels of thinking in biology from that of the layman to that of the most profound theoretician. Our common names very often correspond roughly to body plans (beetle, snail, spider), though it is only fair to point out that in some cases common names relate to taxonomically 'superficial' characters (snails *v.* slugs). At the other end of the spectrum, we find one of the most profound thinkers in the history of biology – the Scot D'Arcy Wentworth Thompson – telling us that the living world is characterized by fundamentally different types of design. That D'Arcy Thompson felt this way is particularly significant, for it was he who devised the most elegant way of looking at *quantitative* differences at lower taxonomic levels (the geometric method called transformation). Despite the success of his method, D'Arcy

Thompson realized its limitations. He states*: 'Our geometrical analogies weigh heavily against Darwin's conception of endless small continuous variations; they help to show that discontinuous variations are a natural thing, that "mutations" – or sudden changes, greater or less – are bound to have taken place, and new "types" to have arisen, now and then.'

We shall take the idea of a body plan for granted from here on, despite its difficulties, since these are so greatly outweighed by its advantages, and shall leave the critics to mutter dark things to themselves. (Book reviews are an excellent source of these mutterings.) An important question, then, and arguably the key question of mega-evolution, is: how do body plans originate? Of particular interest here is whether the 'conventional' mechanism, namely Darwinian selection compounded over vast stretches of time, will suffice – or whether we need to invoke some novel evolutionary process. I shall approach this issue in two ways: first, by considering the overall pattern of mega-evolution, in terms of evolution of the morphogenetic tree, and examining what this tells us about the origin of body plans; second, by looking directly at the question of how body plans first appear, and discussing a novel mechanism which may (or may not) be involved.

Evolution of the morphogenetic tree

Although the group of present-day biologists most closely associated with the Darwinian tradition (population geneticists) are largely unreceptive to evolutionary ideas that are in any way 'developmental', the morphogenetic tree concept meshes in rather well with population genetics theory. What I want to do in this section is to explain this 'meshing-in', at least in outline. To do this, it will be necessary to re-formulate the morphogenetic tree in yet another way. I apologize to those readers who feel that it has already been presented in quite enough different ways in Chapter 10. However, the effort of absorbing yet another picture of the 'tree' now will be repaid by a clearer understanding of the evolutionary discussion that follows.

I want, at this stage, to focus on the *genes* that contribute to

* In *On Growth and Form* (Cambridge University Press, 1917).

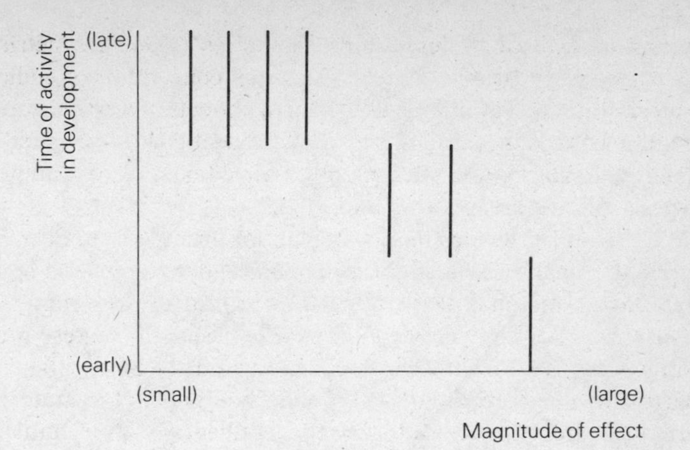

Figure 27. A gene-based version of the morphogenetic tree. Each bar represents a particular D-gene. As development proceeds, the number of active D-genes increases but their average magnitude of effect decreases.

the morphogenetic tree, that is, the D-genes, as described earlier. I also want to bring in the idea of the magnitude of a D-gene's effect on the overall developmental process. The main idea here is simply that genes controlling *early* developmental processes have, on average, *larger* effects than those controlling later ones. This idea is hardly controversial: a gene controlling the formation of the little finger in man obviously has a much smaller effect on overall development than an earlier-acting one controlling, for example, the broad layout of the embryonic body into head, trunk and limbs. In general, the later the developmental stage we are considering, the smaller the effect of D-genes controlling that stage.

This idea of the magnitude of a D-gene's effect can be combined with the earlier idea (p. 128) that the number of active D-genes increases as development proceeds, to give the picture shown in Figure 27, which depicts the genetic contribution to development as a sort of 'hierarchical cascade' process. Only one remaining complication now needs to be considered. This is that in practice we get an idea of a gene's 'magnitude of effect' by what happens when the gene concerned is defective. Yet each D-gene (or D-

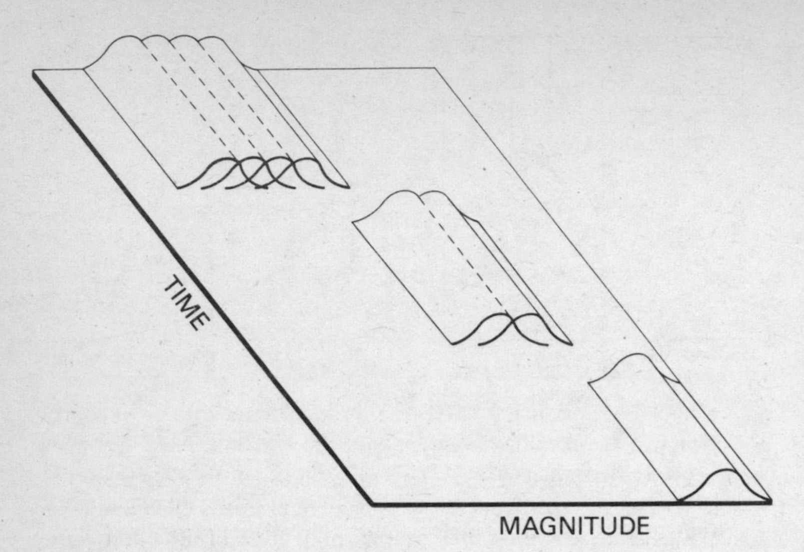

Figure 28. A mutation-based version of the morphogenetic tree. The tree is now three-dimensional. The third dimension, projecting upwards from the time/magnitude plane, is frequency.

locus, to be more precise) is subject to mutations that are themselves varied in the magnitude of defect they produce. This is taken into account in Figure 28, where each gene is represented by a *distribution* of sizes of effect rather than just a single size.

What Figure 28 constitutes, in fact, is a version of the morphogenetic tree that is based on D-*mutations* rather than D-*loci*. This is just what we want, for evolutionary purposes, since evolution is brought about by the incorporation of mutations into the population.

A problem of mutations with large effects on development is that they are usually selectively disadvantageous The general connection between 'magnitude of effect' and 'probability of being selectively advantageous' is shown in Figure 29.* Here we concentrate on two morphological characters only, so an organism can be represented as a dot in two dimensions (very different dimen-

* See R. A. Fisher's *Genetical Theory of Natural Selection* (Oxford University Press, 1930).

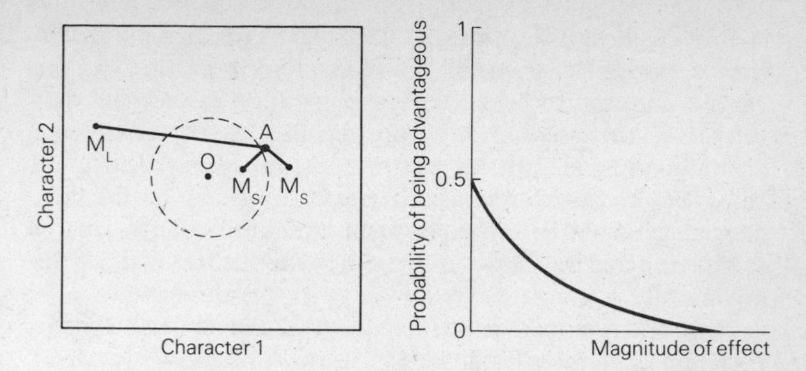

Figure 29. The relationship between a mutation's magnitude of effect and its probability of being advantageous. *Left*: The contrast of mutations with 'small' and 'large' effects. (Fitness decreases with distance from O. The dotted circle round O is a fitness 'isobar' connecting all hypothetical organisms whose fitness is equal to that of the actual organism A.)

sions from the ecologist's spatial ones discussed in Chapter 5!). If organisms of the species we are considering are currently at point A (for actual), while the optimal phenotype with respect to these two characters is represented by point O, mutations (M) producing a point closer to O than A was are advantageous, while those producing a point further away from O are disadvantageous. It only takes a little while thinking about this system to appreciate that, in general, large-effect mutations (M_L) have a lower probability of being selectively advantageous than small-effect ones (M_S). Indeed, mutations with very small effects have a 50 per cent chance of being selectively advantageous, because any very small segment of the circle around O shown is effectively a straight line. Mutations with very large effects (such that the distance AM exceeds the distance 2AO) are *never* advantageous, since they necessarily move the organism further away from O.

We are now in a position to describe one of the two ways in which morphogenetic trees evolve. I shall call this kind of change *phase change* to distinguish it from the other kind, *structural change*, which will be discussed later. In phase change, the structure of the morphogenetic tree remains intact, but the nature of particular links changes owing to allele-replacement at the D-loci controlling those links. An example of phase change is the switch from dextral to sinistral snail shells, mentioned earlier. Here, the particular causative link from one generation to the next that determines the direction of coiling remains in existence after a dextral-to-sinistral switch, but the *nature of the instruction* embodied in that link has changed. (Abolition of the link altogether would be an example of structural change, and it should be noted that structure here refers to the tree and not the organism!)

There are three main points to note in relation to phase-change evolution of morphogenetic trees:

1. More D-genes control later developmental stages, so mutation affecting these stages will occur more often than those affecting earlier ones.
2. A higher proportion of the mutations that do occur will be selectively advantageous at later stages.
3. On the relatively rare occasions where advantageous mutations do turn up at early developmental stages, the adult phenotype will change by a larger amount.

These three points, taken together, give us a general picture of phase-change evolution of morphogenetic trees. What happens is that most of the time there is a 'tickover' of minor changes in late development due to allele-replacement at late-acting D-loci. Occasionally, we get a more radical change due to allele-replacement at an earlier-acting gene.

This is in a broad sense a 'punctuational' view of evolution, though the time-scale involved is much longer than in the 'punctuated equilibrium' theory of the origin of species, and the punctuations or 'jumps' are separated by periods of slow change rather than stasis. Because it is punctuational in a general sense (or saltational, if you prefer), this view of evolution is in contrast to strict Darwinian evolution where, as Darwin kept repeating, '*Natura non facit saltum*' ('Evolution does not make jumps'). Since I first proposed this theory in detail in 1984,* some population geneticists, viewing themselves no doubt as defenders of Darwinism, have objected to it because of this apparently anti-Dar-

* In *Mechanisms of Morphological Evolution*, published by John Wiley.

winian stance. Yet to my mind it is a decidedly pro-Darwinian theory, both because the only mechanism involved is standard Darwinian selection and because *most* (but not all) changes are minor ones.

In the end, this view of evolution must be assessed in its own right, and whether it is better considered as pro- or anti-Darwinian is relatively unimportant. I sometimes wonder how the most blinkered of the neo-Darwinians see long-term morphological evolution. Their views are rarely exposed since they spend most of their time attacking alternative views, but several possibilities are open to them. One is that the morphogenetic tree idea is entirely wrong. A second is that only the top of the tree, above some arbitrary cut-off point, evolves. A third is that the distributions of Figure 28 all extend back to the zero-point on the magnitude axis, which would allow evolution to act on minor variants at *all* developmental stages, and perhaps cause it to be blind to the difference in the *peaks* of the distributions. I doubt if any of these views is correct, though the third option has more to recommend it than the first two.

We now turn to the second kind of evolution which occurs in morphogenetic trees, namely *structural change*. Here, the actual pattern of interconnection of causal links that constitutes the tree itself changes. No doubt, through the whole history of life, many different kinds of structural change have occurred in many different lineages. I want to concentrate on what I believe is the most important general structural change that has taken place, which is basically 'growth' of the tree.

It is a widely acknowledged fact that, in the long term, evolution has resulted in increasingly complex kinds of organism. Animals started off unicellular, then became jelly-like or worm-like, and finally ascended to the level of complexity seen in molluscs, arthropods and vertebrates. A parallel progression in complexity occurred in plants. As usual there are exceptions, with a decreasing complexity of parasitic forms being the most often-cited example. But in general, evolution has tended in the direction of increasing complexity of *organismic structure*. This increase must be underlain by increasing complexity of the developmental

process, that is, by increasingly complex morphogenetic trees.
The general point I am making, then, is that while they undergo phase change by allele-replacement, morphogenetic trees also grow, probably most often by 'terminal accretion', since the insertion of new developmental stages in between previously existing ones would be fraught with problems. The genetic basis of *growth* of morphogenetic trees could be gene duplication and divergence (a well-known phenomenon), but I doubt if this is the only mechanism. Whatever the mechanism at the molecular level, the reason we get increasingly complex morphogenetic trees and hence increasingly complex organisms may be that the probability of success of mutations deleting parts of a well-adapted organism is lower than those adding parts; consequently the net change is in the upwards direction.

We can now picture the evolution of morphogenetic trees, involving both phase change and structural change, by presenting a particular tree, representing the development of an organism at a particular point in time, and a second tree representing one of its descendants many millions of years later. This is done in Figure 30, where a switch from a solid to a dashed line represents allele-replacement. As can be seen, the later organism contains more phase changes in later than in earlier stages of the original tree; and its tree is also larger, with the most recently evolved 'top branches' having appeared during the intervening period. (A similar comparison, using heterogeneities rather than genes, and based on the picture of the morphogenetic tree given in Figure 17, could easily be made.)

This pattern of change would explain 'Von Baer's Law', which states, essentially, that early stages of development have more in common, between two taxa, than later ones. This law is in contrast to the view, usually attributed to Ernst Haeckel, that the *adult* stages of ancestral organisms appear in the developmental process of their descendants. The latter would only hold, in the morphogenetic tree picture, if structural change occurred without phase change, and if *all* structural change, rather than just some of it, was towards growth. This is a pleasing outcome, since we now know that Von Baer was right and Haeckel wrong, but as usual we must be cautious about accepting compatibility of theories as evidence of their correctness.

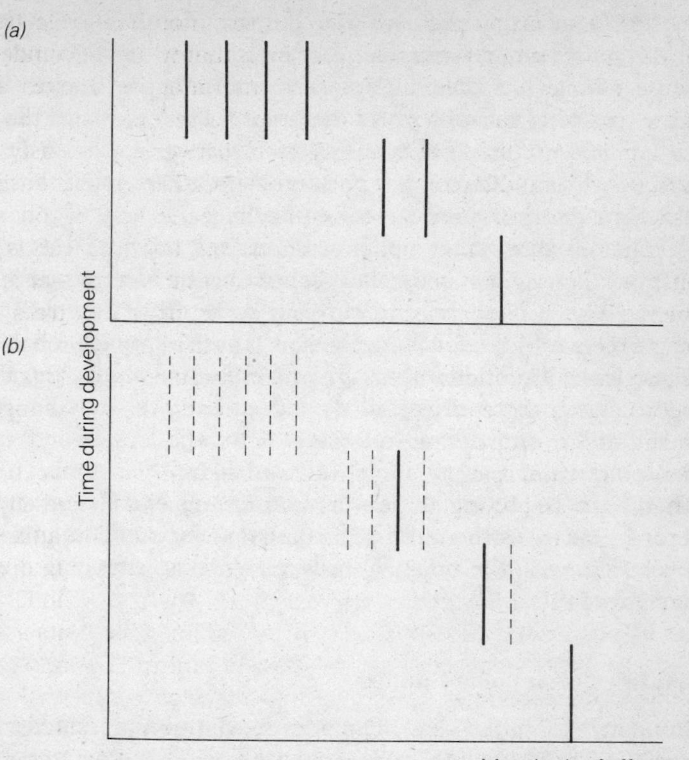

Figure 30. The evolution of the morphogenetic tree. (*a*) Ancestral organism. (*b*) Descendant, many millions of years later. The dashed lines indicate new alleles or loci. In (*b*), the developmental period has lengthened, and the tree has grown by extension to the right, not to the left.

Finally, we turn to the question of what this general picture of long-term morphological evolution tells us of the origin of body plans. The answer is, I think, that it tells us very little except that body plans *could* originate in a conventional way at least in as much as we need only invoke the standard selective process coupled with extinction of intermediate forms. Admittedly there is a temptation to see the occasional allele-replacement at an early acting D-locus as being associated with (though not the *whole* cause of) the origin of a new type. But even then, although this is in a sense unconventional, the mechanism involved at the population level is still standard Darwinian selection. Incidentally, an example of this process is the origin of the gastropod family Clausilliidae, which is almost exclusively sinistral, from a dextral ancestor by mutation of, and presumably standard selection upon, the gene governing the direction of coiling.

Whether this rather negative statement provides the whole story of the origin of body plans cannot yet be ascertained. However, I would like briefly to consider some other approaches to this problem in the following section. These approaches are in a sense less conventional than the one presented above; yet, if the mechanisms to be discussed do indeed operate, they do so in addition to, and not as alternatives to, the broad pattern of morphogenetic tree evolution presented so far. Also, I must admit that I am beginning to see the continuing elucidation of this broad pattern as the most important task for students of mega-evolution, with the origin of body plans being simply one component of it.

On large evolutionary jumps

Fundamental to the view of long-term evolution presented in the previous section is the picture of a mutation's declining probability of improving the organism as its magnitude of developmental effect increases, as illustrated in Figure 29. Early-acting D-genes have large effects partly because of their own direct contribution to development, but partly also because of their role in switching on or off batches of later-acting genes. If the early gene mutates, it is likely to be put out of tune with its later 'modifier genes', and so to give rise to a phenotype that will be selected against. Clearly, the more 'modifiers' a particular D-gene has, that is the further 'down' the tree it is, the greater the problem. This phenomenon of evolutionary inertia of early-acting genes because of problems such evolution causes for the cascade of later developmental processes has been referred to as 'generative entrenchment' by the American philosopher of biology William Wimsatt. This seems a useful label. One consequence of the pattern of evolution described in the last section, where the predominant structural change in morphogenetic trees is growth, is that most D-genes will become more generatively entrenched as evolution proceeds. As they have progressively more subsequent processes thrust upon them, their evolutionary inertia is bound to increase.

A question which begins to appear at this stage is whether D-genes in the 'trunk' of the morphogenetic tree of a fairly complex organism can evolve at all. In other words, have they become so generatively entrenched that all mutations in them would be disadvantageous, as in the case of the large-effect mutations shown in Figure 29? One cannot provide a concrete answer to this question, but clearly a problem of this sort *may* exist. I now want to inquire whether – supposing these key D-genes to be incapable of conventional evolution for this reason – they have any escape route from evolutionary stasis. That is, are there any unconventional mechanisms by which they might evolve, despite their high degree of generative entrenchment?

There are two possibilities here, the only thing in common between them being that they are both based on the pattern shown in Figure 31 (contrast with Figure 29). Here, the probability of a mutation being advantageous first declines, *but then increases*, as we go in the direction of increasingly large developmental effect.

The first reason why this may happen is most easily seen if we think of the magnitude of effect being expressed as the number of separate phenotypic characters that are altered by the mutation. Consider again the example of chirality in snails. Imagine a snail with a 'reversed' shell, but without the muscle attaching the body to the shell being similarly reversed; clearly, such a design would fail. The reason that mutations of the chirality gene are not always hopeless disasters is that they change *all* characters in the same way, and therefore retain a high degree of *coadaptation* of the whole organism. Theoretical considerations of this sort might almost suggest a symmetrical U-shaped curve for Figure 31, with a minimum at the position where 50 per cent of characters had been altered. However, it is clear that this is not the case, and that in reality the peak at the right-hand side is very much smaller than that on the left.

In the situation just described, some large-effect mutations escape from the evolutionary inertia that one might have predicted for them not by a novel mechanism, but by an 'old' one (Darwinian selection) acting in an unexpected way. How often early-acting D-genes with major effects will find this particular escape route available to them is impossible to determine, but I would guess that it might not be very often. This is not a problem, though, as major changes occur only rarely in evolution.

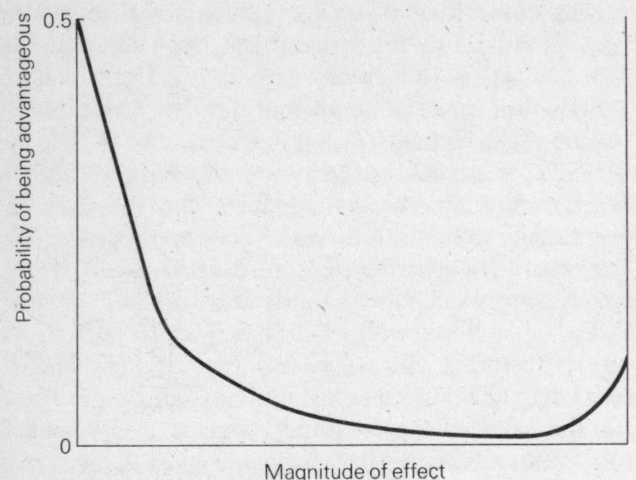

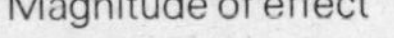

Figure 31. A revised picture of the relationship between a mutation's magnitude of effect and its probability of being advantageous.

As already noted, the second 'unconventional' mechanism also involves the background provided by Figure 31. Now, however, the subsidiary peak at the right-hand side occurs through a different, and novel, mechanism, which requires some explanation. It will be better, this time, if we think of 'magnitude of effect' not as the *number* of characters affected but as the size of effect on one particular major structure. Let us consider, as an example of such a structure, the segment of a primitive arthropod bearing the mouth-parts. A mutation in a D-gene controlling the early development of this segment will have a large effect on the developmental process and hence on the adult phenotype. Mutations of this sort will tend to have two effects. First, they will decrease the degree of adaptation towards the acquisition of the type of food our hypothetical insect normally consumes; second, they will decrease the degree of coadaptation of the organism as a whole (for example by making the mouth-parts too large for that segment to be easily supported by the legs).

We are accustomed to believing that such a mutation will rapidly disappear as a result of standard Darwinian selection. However, this need not necessarily occur unless the mutation is actually lethal (or sterility inducing). Why not? Basically, because the mutant organism may be able to utilize a different type of food, *and hence have its population regulated by a different resource*, compared with the original organism. In this case, it ceases to be meaningful to measure fitness in terms of competitive ability, as noted in Chapter 11. Rather, we now measure fitness in the 'absolute' sense of whether the new organism can build up an (independent) population.

The main problem facing such a mutant organism concerns breeding. Mutations are thought of as rare events. If a single mutant organism appears in the population, it may have one of two problems. If the mutation has also affected its reproductive compatibility, so that it is now isolated from its parent-form and can only breed with other mutants, it finds itself with no mates. If its reproductive capability is unaltered it has plenty of mates, but these are of the 'old body plan', and we have the odd situation of one species with two discretely different structural designs within it.

Whether these problems are insurmountable is difficult to tell, though some possible solutions suggest themselves. The 'mateless mutant' will not be mateless if all his siblings are mutant, as they will be if the gene concerned exhibits 'delayed Mendelian inheritance' as described in Chapter 7. And it is noteworthy that the most major-effect D-genes at the base of the morphogenetic tree are inherited in just such a way. The 'two-body-plans' situation may be unstable and lead to subsequent reproductive isolation of the two types – though exactly how this would happen is difficult to envisage.

I have called the process of 'viability selection' in a different niche 'n-selection' to distinguish it from Darwin's competitive selection, which I labelled 'w-selection'. (The use of these particular letters is of little consequence – an explanation of why n and w are used would take up much space and reveal very little!) It still seems appropriate to use the term selection, because some large-effect mutations causing a niche-shift sufficient to render competitive ability irrelevant will fail and others will succeed – so there is a kind of selection going on.

As we have seen, n-selection is beset by a number of problems, and is at most likely to be a very rare process. Are there any advantages in invoking it at all? There would appear to be two. First, n-selection will only work when there are few other species, because if there were many the evolutionary jump involved might still take the mutant out of its parent-species' niche space, but would probably take it into the niche of some other species, which would mean that competitive ability was again important. New body plans originating by n-selection would thus only appear *early* in the evolution of multicellular organisms, because only then would there be substantial amounts of unutilized niche space into which to 'jump'. We know that most new body plans did originate early in multicellular evolution (in the period 800–500 MY ago) during a phase known as the 'Cambrian explosion'. Since then, few major new body plans have arisen. So origination through n-selection has the advantage of tying in with this fact. However, an alternative explanation of the lack of recent body-plan originations lies in the increasing generative entrenchment of key early-acting D-genes, and this explanation could apply to *conventional* origin of body plans through competitive Darwinian selection.

The second advantage of invoking the occasional instance of n-selection is that by doing so we can produce an interesting picture of the way the level of coadaptation might change during evolution. This can be best seen by contrasting it with the coadaptational picture when only Darwinian selection is permitted. In either case, I will assume, as seems reasonable, that the presence of a particular allele at any particular D-locus determines the optimal allele at all subsequent ('modifier') D-loci. If we start with a well-coadapted organism, allele-replacement at an early-acting D-locus will occur under Darwinian selection only if the new

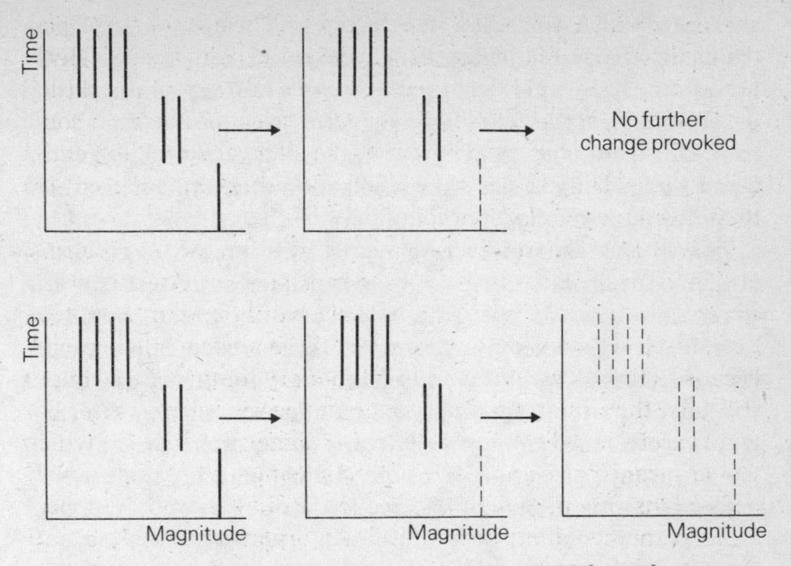

Figure 32. Phase change in morphogenetic trees. *Top*: The replacement of an allele at a major locus by Darwinian selection retains (or improves) the level of coadaptation. *Bottom*: Allele-replacement by n-selection (middle) decreases coadaptation and stimulates phase change further up the tree.

allele is even more 'in tune' with its modifiers than the old allele was. Thus the replacement at the early locus sets up no pressure for subsequent change; the organism remains well coadapted, and even improves slightly in this respect. In contrast, an allele-replacement at an early locus that becomes established through n-selection will *decrease* the level of coadaptation, and set up a strong pressure for further allele-replacements at later-acting D-loci that will allow the organism to regain the level of coadaptation that it has lost. These two possibilities are shown in diagrammatic form in Figure 32. At this stage in our knowledge of coadaptation (a notoriously difficult concept to quantify), it is impossible to say which picture is nearer to the truth.

The difference between the 'conventional' picture of morphogenetic tree evolution presented in the last section, and the picture of waves of re-coadaptation sweeping upwards through the tree driven by occasional instances of n-selection in its trunk discussed above, is a subtle one. Evidence could only be easily brought to bear on which view was correct if we could directly examine *early* stages of evolution, which we can never do. That being the case, the 'conventionalist' will say that there is no direct evidence for n-selection, and so the alternative view must be correct. But equally one can argue that there is no *direct* evidence for a Darwinian origin of a body plan – black *Biston betularia* certainly do not constitute one! Thus in the end we have to admit that we do not really know how body plans originate. This, like species-level punctuations, is readily seized upon by the creationists, who would love to see evolutionary theory as a whole fall flat on its face. Evolutionists, on the other hand, see their ignorance on the origin of body plans as an interesting problem for the future rather than as an indication that their whole approach to the history of life is erroneous.

Summary

Discussions of macro- and mega-evolution are particularly prone to misinterpretation and to unnecessarily heated debate. A brief summary may help to prevent such a fate befalling the discussion presented in this chapter. This is given below, in the form of a list of the most important points.

1. It is more helpful to think of three levels/time-scales of evolution (micro, macro, mega) than just two, even though it is not possible to draw precise boundaries between them.
2. At the level of macro-evolution, a punctuational pattern is commonly but not universally found in the fossil record. Various mechanisms might underlie this pattern, and some of these are compatible with conventional Darwinian theory.
3. At the level of mega-evolution, one can concentrate either on the origin of body plans or on formulating an overall scheme for the evolution of development, of which the origin of body plans is one component. The second of these strategies is preferable.
4. Long-term evolution of development can best be approached by combining a model of development (the morphogenetic tree)

with a body of theory on the fate of mutations in populations (neo-Darwinism). Such a combination can explain Von Baer's law and the evolution of morphological complexity. It also provides some clues on the origin of body plans. Evolution of the morphogenetic tree involves two distinct components, one called *phase change*, the other *structural change*.

5. The origin of body plans may involve mutation of key genes controlling early developmental processes. Whether this occurs through standard Darwinian selection or by a novel process called n-selection is not clear; but the two mechanisms have very different consequences for the way in which organismic coadaptation varies throughout the evolutionary process.

One final point. The Darwinian and developmental approaches to long-term morphological evolution are not *alternatives* to each other, though they often appear so. Darwinism concentrates on what happens in populations, and says little about the nature of mutational processes occurring in *individuals*, even though the existence of these is a pre-requisite for population-level changes. Some Darwinists go so far as to write off mutations as random, though it is doubtful if they are truly random in any sense. The early use of 'random' could perhaps be defended by the need to provide a contrast with Lamarck, where mutations were directed to the needs of the organism. Since Lamarckism is dead, we need no longer foist a false randomness on mutations. 'Developmental evolutionists' are guilty of a different sin, and in a sense the opposite one. They often concentrate on mutational changes in the developmental processes of *individuals* without considering the fate of such changes in populations, particularly the fact that most interesting developmental mutants are rapidly *removed*, rather than fixed, by natural selection. We will never arrive at a completely satisfying theory of the evolution of development until this false conflict is stopped in its tracks. Selectively driven evolution of the morphogenetic tree, involving as it does both phase change and structural change, is my contribution to a combined Darwinian/developmental approach. No doubt the version given here is incomplete, and inaccurate in many details, but I hope it is a step in the right direction.

13 TOWARDS A UNIFIED THEORY?

Some physicists spend most of their time searching for a GUT. This is not because they are thwarted upper-case biologists; a GUT is not a lowly digestive tract but rather a Grand Unified Theory. One of the aims of physics is to arrive at such a theory, which will subsume all other physical theories and perhaps even render further pure research in physics redundant. The GUT is the pinnacle of scientific endeavour – maximum generality and simplicity: everything explained by a single, all-embracing theory.

It should be clear by now that we have no biological GUT. If we had, this book might only have been a single chapter long! Whether biologists can realistically aspire to such a long-term aim is uncertain; one can take two views of the situation. In one view, sciences such as biology and geology are the poor relations of the grand sciences of physics and chemistry. The latter are applicable to laws that operate throughout the universe, while the former are relatively low-key studies of local phenomena which will never develop into anything very general or profound. The alternative view is that biology can transcend its biospheric origins and develop principles that will apply to life wherever we encounter it. The transcendent generalities are still undetermined, of course, but principles of self-organization and homeostatic regulation are good candidates. In this view, theoretical biology is only barely separated from cybernetics.

The truth may turn out to be somewhere between these two extreme views, though my feeling is that it will be nearer the latter, more optimistic one. If so, it may not be too unrealistic to hope to see, one day, a biological GUT. However, since that day has not yet come, there is little point in devoting much space to speculation about the precise form our GUT will take. My concern, below, is with the more immediate future.

Back to order

Although we have no single 'grand' or 'general' *theory* of biology, we do have the general *theme* of order, with which this book started. The word 'order' has not been mentioned much since Chapter 1, but this does not mean that in writing Chapter 2 I changed my mind about its importance! Rather, when discussing specific aspects of order in biological systems, one uses the more specific terminology that is relevant to the topic at hand. All the individual theories discussed in Chapters 5 to 12 involved ordering mechanisms, but to make progress in studying them we had to go beyond the fact that an ordering mechanism was involved and inquire into the *kind* of ordering mechanism with which we were dealing.

Some authors like to make distinctions between the concepts of order, organization and complexity. I am not sure whether these distinctions are particularly helpful, at least in the context of this book. The meanings of the three terms are certainly very closely related, though one can construct scenarios in which differences begin to appear. For example, it could be argued that a house is more complex than a bungalow, but no more ordered. On the other hand, a recently bulldozed house is much less ordered than a still-standing one, but it could perhaps be argued that it is equally complex! If you choose to recognize distinctions between order, organization and complexity, I think you have to recognize that biological systems tend to increase in all these things.

The distinction that is important here is one that was introduced in the first chapter – namely the distinction between mechanisms that increase order/complexity and those that maintain it. The processes dealt with in this book (and others) tend to fall into one or the other of these categories. Weismannism, Mendelian inheritance and population regulation are all mechanisms of *maintenance*; evolution, development and ecological succession (the last of which I have not dealt with, since we don't really understand it yet) are all mechanisms that *increase* order or complexity.

It is tempting to speculate that *en route* to our single unified theory we shall reach a stage where the number of theories has been reduced to two: a theory of ordering or self-organization processes, and a theory of regulatory processes which in some way maintain the *status quo*. These are the very two things mentioned above as possible 'transcendent generalities' that will apply to extra-biospheric living systems if and when we find them. However, we are still jumping ahead. Our more immediate concern should be the compatibility of the *several* theories that currently exist.

The biological citadel

In Chapter 4, I referred to the theoretical superstructure on which we hang our facts as the biologist's citadel. This was displayed as a group of interconnected areas of theory, in Figure 7. We have now got to the point where we can climb out of the citadel again, having looked around its various rooms, and can inspect it once more in its entirety. Since we are all now qualified surveyors, we should be able to give it some sort of report, and to conclude whether or not it is in reasonably good shape.

The main questions here, as I have repeatedly stressed, are: first, whether each theory is in tune with the relevant parts of reality; and second, whether the various 'neighbouring' theories are in tune with each other. As I have said, neither of these compatibilities guarantees the correctness of our theories, though if we find them it is certainly a good sign. But *in*compatibilities of either sort do indicate that something is wrong, even if it is not immediately clear *what*.

The connection of each individual theory with reality has been commented upon on the way through, but the following brief summary of this kind of compatibility will help. Weismann was basically on the right lines in his anti-Lamarckian stance, though the germ plasm is not just quite as independent and continuous as he might have liked. Darwin's natural selection is now a well-established evolutionary mechanism, though doubts have been cast on its applicability to very minor changes (where Kimura's 'neutralist' theory has much to recommend it) and to very major ones (in particular the origin of higher taxa). Mendelism is the predominant kind of inheritance in sexually reproducing organisms, and all inheritance is particulate. The central dogma of molecular biology has held out, despite the subversive activities of retroviruses, and provides us with a molecular counterpart to Weismannism. Cells differentiate almost exclusively through variable activity of genes rather than by permanent loss of those genes that are unnecessary. The diversificationary aspect of development, which includes both cell differentiation and pattern formation, may be underlain by a hierarchical structure of causative links – the morphogenetic tree. This proposition, to which the present book makes an original contribution, remains relatively untested. Most populations in nature are regulated around an equilibrium due to the action of the so-called 'density-dependent factors' of resources and predators. In the long term, later developmental stages of organisms evolve faster than earlier ones, as maintained by Von Baer. One possible explanation of this arises from a study of the way in which conventional Darwinian selection works against a background of the morphogenetic tree. The origin of the major groupings we call body plans is still obscure, but seems likely to involve evolution of genes towards the base of the morphogenetic tree, either by conventional 'competitive' selection, or through 'viability selection'.

It emerges from the above brief account, and from the fuller discussion of the preceding chapters, that most of the major biological theories are largely compatible with reality, while others, particularly in the areas of development and long-term evolution, remain relatively untested. Nowhere do we find that a current theory clashes in an important way with the relevant facts. So as regards this sort of compatibility, we have cause for reasonable satisfaction, though there are no doubt some challenges yet to come.

What of compatibility of the theories themselves? Again, we find a broadly satisfactory picture. Figure 33, which is a modified version of Figure 7, shows this picture in outline, with solid lines indicating compatible and important links, dashed lines indicating links that are either problematic or unimportant.

The 'central axis' of Weismann/Mendel/Darwin presents no problems at all. The theories of these three men fit together like the three-way counterpart of a 'lock and key' (whatever it is!). Even if we accept Kimura's theory of neutral evolution at the

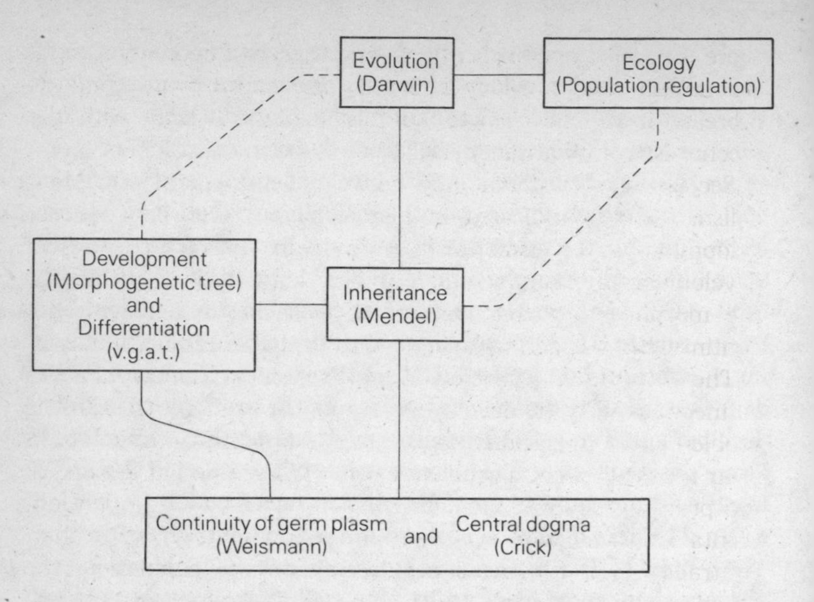

Figure 33. Compatibility scheme for major biological theories (v.g.a.t. = variable gene activity theory).

molecular level, this is just as compatible with Mendelian inheritance as is natural selection. Indeed, predominance of a non-selective mechanism at the molecular level is not even at odds with the predominance of Darwinian selection at the level of organismic evolution, as Kimura himself has frequently pointed out.

The link between selection and population regulation is equally firm. Indeed, as we have seen, Darwinian selection is based on the idea of competing variants sharing the same regulatory factor. Essentially, fitness and competitive ability are the same thing. If this were not so, there is no reason why one variant should go extinct because of the existence of another.

The link between Mendelian inheritance and population regulation has been reduced to a dashed line not because of incompatibility, but because in the end I don't think that much of a connection exists here. Most theories of population regulation will work regardless of the nature of inheritance, and although there is at least one explicitly genetic theory of population regulation, this is not widely believed, and at any rate connects more with genetic *changes* (that is, evolution) than with the mechanism of inheritance itself.

Secure links exist not only between development and Mendelism, as was indicated in Figure 7, but also between development and Weismannism. Admittedly, current theories of development are very tentative, but at least there is nothing in the morphogenetic-tree concept that clashes either with the continuity of the germ plasm or with particulate inheritance.

The only potential problem in our theoretical framework is the connection between development and Darwinism. This link is problematical in various ways – including that it is not even clear to what extent a problem exists! As I stressed at the end of the previous chapter, I personally see Darwinian and developmental approaches to evolution as complementary rather than contradictory, with one concentrating on the population, the other on the individual. We shall never understand evolution if we see no developmental pattern in the occurrence of mutations. Nor shall we understand it if we refuse to recognize that we need somehow to get from mutant individual to mutant population if an evolutionary change is to be achieved. The morphogenetic-tree concept, embodying both a pattern of occurrence of mutations, and a pattern of selection upon them, is intended to be a combined Darwinian/developmental approach. Yet I am well aware that it does not get far enough away from natural selection for many anti-Darwinians to accept it, and it makes too much of the evolution of major genes (even though this is seen as a very rare event) to gain general acceptance in the neo-Darwinian camp. Perhaps in the end I shall persuade them both, but this is a tall order!

Apologia

Apart from the very different views of the living world held by neo-Darwinians and 'developmental evolutionists', there isn't too much to worry about in terms of incompatibility of the various biological theories we have discussed. However, the reader must beware here, because compatibility of theories can be engineered by omitting theories that do not fit the overall picture, just as trends in sets of data can be engineered by the omission of a few numbers that mess up the overall picture the experimenter would like to see. Clearly, I haven't included in this book all those ideas in biology that have at some time or other been labelled theories or laws. To have done so would have required a vast volume that you would probably never have read! But, while this is a reasonable excuse, you need to be reassured, I think, that my policy has been to omit relatively unimportant theories rather than to omit those theories that 'don't fit'. This section takes the form of this assurance (with a few complications) coupled with an apology to those who have spent many years working on theories I have left out.

In fact, the omitted theories fall into several categories. First, there are 'old' major theories that are so thoroughly accepted that we tend to take them for granted – a fate that is fast approaching Mendelism. A prime example of a theory of this kind that I have left out is the 'cell theory', which proclaims that living organisms are made up of small units called cells. It is now such a commonplace observation that this is true that it seems odd to think of it as a theory any more. (This 'senility' stage is a part of the history of theories that I didn't go into in Chapter 2.) Of course, being taken for granted and being universally true are not the same thing and, as with other biological theories, we find exceptions – for instance the syncytial (non-cellular) slime moulds. However, the cell theory is both *generally* correct and compatible with all the major theories we have discussed.

Second, I have omitted theories that seem to be correct but are of relatively minor importance in biological thought. An example of this is Bergmann's rule, which says that warm-blooded vertebrates get larger as the climate gets cooler – by going north or south away from the equator, for instance. Harvey's theory of circulation of the blood in man is an example of a theory which is both 'minor' and old enough to be taken for granted. No theory of very restricted applicability, such as these, is likely to cause us to question the 'major' theories of particulate inheritance, natural selection and so on.

I have to admit that I have omitted some aspects of evolutionary theory that I would *not* describe as unimportant. I have covered what seem to me to be the *most* important aspects of evolution, and have then guillotined the rest of this area at some arbitrary point in the spectrum of declining importance, as I perceive it. One reason for this is simply that a book on theoretical biology could very easily become a book on evolution, and there are plenty of those about already. However, I would like to mention just one omitted evolutionary topic before moving on. This is the business of evolution by altered *timing* of developmental processes – something that goes by the name of heterochrony (which includes neoteny, paedogenesis and other processes with equally horrendous titles). I am firmly convinced that by invoking a third type of change – *distortional change* – heterochrony can be accommodated within, and seen as part of, the idea of morphogenetic-tree evolution. This could be pictured by accelerating or delaying some of the tree's 'branches' relative to others, which is fundamentally different from the other two types of evolutionary change (*phase* and *structural*) that occur in morphogenetic trees. Two further comments on heterochrony: first, it seems most relevant to the *macro* level of evolution, and an example of its occurrence is in the evolutionary divergence of man and chimp; second, it cannot operate in isolation from the mutational creation of novelty – you can produce only a rather limited set of organisms by varying the relative timing of a fixed array of developmental processes. Finally, anyone unhappy with my omission of heterochrony can take comfort from the existence of a recent book devoted largely to this subject.*

One continuing major controversy that I have omitted is in the area of taxonomy, and involves the modern 'cladist' school, which owes its origins to the German palaeontologist Willi Hennig, and their opponents, the more conventional school of 'evolutionary taxonomists'. I have omitted this debate for two reasons, neither of them particularly good ones. First, I think the general reader finds taxonomy an intrinsically boring subject. Second, I find the whole argument rather annoying because it is not about evolutionary mechanisms, and yet some cladists act as if it were, and

* *Ontogeny and Phylogeny*, by S. J. Gould (Harvard University Press, 1977).

argue (for no good reason that I can see) for the instant dismissal of Darwinism. For example, a leading cladist, Donn Rosen (writing in the book by Pollard – see Suggestions for Further Reading), proclaims that there is 'no need to placate the ghost of neo-Darwinism; it will not haunt evolutionary theory much longer'. If, as I suspect, these utterances come only from the 'lunatic fringe' of the cladist movement, then they can be written off as black sheep theories (see below), leaving the rest of cladism to be incorporated into the biological citadel – preferably in its original, Hennigian, form.

Finally, I have omitted theories that are directly contradictory to the mainstream of theoretical biology when these are held by a few individuals and rejected by most of the biological community. I label these, rather unfairly perhaps, as black sheep theories. Examples of this include E. J. Steele's neo-Lamarckian theory (mentioned briefly in Chapter 5) and a theory due to E. O. Wiley and D. R. Brooks that says, contrary to the conventional wisdom, that evolution proceeds towards increased entropy (or dis-organization). The supreme example is Rupert Shelldrake's recent theory of 'morphic resonance'. The reader who has not come across this bizarre theory is more fortunate than those of us who have, and I will not seek to unenlighten him! Sometimes black sheep theories later gain acceptance, and it turns out to be the white sheep that were wrong, despite their number; but more often they remain estranged from the mainstream and eventually disappear.

Now it could be argued that I have done exactly what I hoped not to do – falsely engineered a compatibility within theoretical biology by omission of the black sheep theories. It is certainly true that these are, by their very nature, *not* compatible with orthodox views. But those who argue in this way are expecting something that can never be achieved – complete homogeneity of human thought within a particular area of study. However long science endures, there will *always* be individuals and small groups holding to a 'heterodox' viewpoint. Mainstream science is a bit like a majority verdict from a jury. This of course raises the question of what sort of majority is necessary for a theory to be accepted within the prevailing conceptual framework. This seems to me a major question in the sociology and history of science, and one that has not received the attention it deserves. I could not begin to answer the question, but I do know one thing: prestige counts (unfortunately) for a lot in science, and the scientific jury is a long way from 'one man one vote'.

Life-cycle theory

Going back to the idea of a general theme of theoretical biology, order might be objected to because of course the abiotic world is not entirely unordered, and so ordering processes do not uniquely apply to the biosphere – even though they are particularly evident there. Another general theme exists, however, that *is* unique to living organisms, namely life-cycles. I would like to argue that the life-cycle could perhaps represent a focal point for theoretical biology, even to the extent that this book could have been entitled *Life-Cycle Theory*. Since most of us first met the idea of life-cycles at primary school in the context of frogspawn, tadpoles and frogs, this may seem like a ludicrous elevation of a very elementary phenomenon of natural history to a level of theoretical importance that it does not deserve. However, if you feel this way, try to suspend your disbelief for a moment!

The first observation we should make here is that an organism does not *have* a life-cycle – rather, it *is* a life-cycle. It is an entity of four dimensions, not three. A horse is a horse from fertilization till death, even though the small ball of cells that represents the horse at a very early stage of development would hardly be recognized as such if presented in isolation to a perplexed observer. The same applies to any other organism.

The second observation is that the entire history of life in the biosphere (covering a period approaching four billion years) could be represented by a network of interconnected life-cycles. Every living organism (or life-cycle) is connected to every other, to every extinct organism, and to the 'first organism' (if one could recognize it) by a bridge of many life-cycles.

If we focus on any small part of the network of life-cycles, we find that some biological processes (development, ageing) provide the parts of a single life-cycle. Others (inheritance, Weismannism) provide the links between consecutive life-cycles. Evolution charts the spread of some types of life-cycle at the expense of others. In other words, all the processes about which this book has theorized can be seen in a life-cycle context; and it could be argued that they are seen more clearly this way than any other, as it emphasizes the true four-dimensional nature of the phenomena we seek to explain.

One, two or three biologies?

It should be clear by now that my focus throughout has been on organisms (or life-cycles) rather than on any other level in the biological hierarchy. True, we have descended to the level of the cell (and even occasionally to DNA) in an attempt to understand the organism's development, and we have ascended to the population level to look at theories about the dynamic behaviour of *groups* of organisms. But we have had little to say about most molecular phenomena or about processes such as nutrient cycling operating at the level of whole ecosystems.

I call my organism-centred approach 'Middle Biology'. Above this, we have 'Upper Biology', or community and ecosystem ecology, if you prefer; and below it 'Lower Biology', consisting of biochemistry and biophysics. Of course, the distinctions are not particularly clear-cut, and might be objected to for all sorts of reasons. Nevertheless, these distinctions do seem to indicate some sort of discontinuities in biological theory. For example, consider theories I have omitted in the Upper and Lower Biologies. In ecological energetics, there is a 'theory' that energy flow between trophic levels doesn't exceed about 10 per cent. In biochemistry, there is a large body of 'theory' on enzyme kinetics. Neither of these seems to cast light on organism-level phenomena, such as development – despite the fact that without the acquisition of energy from the environment, or the activities of enzymes, development could not proceed.

One possibility, then, is that we shall end up with not one integrated body of biological theory but three. An alternative view (the Two Biologies) is that all organismic phenomena will ultimately be explained in molecular terms and all ecological ones in terms of populations; populations being, in a sense, the molecules of the ecosystem (and individuals its atoms). The most optimistic proposal, of course, is that we shall end up with a single, unified body of biological theory extending from molecule to ecosystem (even if it is a 'body' rather than a single theory or GUT). In one sense, we already have this situation, since there is no clashing of theories from the two (or three) biologies. The remaining problem for aspiring 'monists' is not so much that theories in the different biologies are incompatible, but rather that the links between them are very weak.

The hold of the higher organism

The previous section was intended to emphasize that the reasonably coherent body of biological theory described in this book does not extend throughout all levels of life from molecule to ecosystem. This one is intended to emphasize another restriction to its generality – namely that it tends to concentrate on 'higher organisms'. If the title conjures up a mental image of biological theory receiving a bear-hug, this may not be too inaccurate! Of course, by 'higher organism' I mean to include, as usual, most plants and invertebrates. But this still leaves a large chunk of the living world, to significant parts of which some of the theories described herein do not apply.

I am thinking particularly here of unicellular and sub-cellular organisms, including the bacteria and viruses. Here, diploidy is rare, so Mendelism in its usual form does not apply. And since multicellularity is absent, the concepts of cell differentiation and development are largely meaningless. Furthermore, there is no germ plasm as such. No doubt these organisms still evolve (predominantly by natural selection) and their populations must still be regulated; so they do not escape our theoretical net entirely. But there is no doubt that theoretical biology in its present form is strongly orientated towards the higher organism, in the form of the multicellular, sexually reproducing diploid.

While this message could be taken as a singularly pessimistic one, implying as it does a restriction to the generality of our theories, there is another side to the coin. The higher, more complex organisms have more theories relating to them precisely because they are more ordered and more complex. Since it is order and complexity that theories seek to explain, the dominance of the higher organism in biological theory is inevitable, and is a reflection of the trend towards order and complexity embodied in evolution, which, as we have seen, is itself a central strand in theoretical biology.

14 ATTACKS ON BIOLOGICAL THEORY: CREATIONISTS AND OTHERS

If there is a continuing controversy in a particular area of science, researchers working in that area need to take account of both opposing views, even though they may strongly support one of these views and not the other. Their publications often start by acknowledging the controversy, and go on to provide evidence pointing, they hope, to its resolution in their favoured direction. This is of course a very different approach from omitting any mention of their opponents' view, as if it simply wasn't there. Also, broader scientific publications (such as this book) examining a range of different areas of research 'at a distance' normally try to adopt a balanced approach and to consider both sides of the argument, albeit in less detail than a 'deep and narrow' research article. For example, in Chapter 6 I discussed both 'selectionist' and 'neutralist' views of molecular evolution. Even when a controversy has been settled, examination of the difference between the accepted and the rejected theory can often be more informative than description of the accepted theory in isolation. Hence my approach of contrasting Mendelism with blending inheritance in Chapter 7.

This picture of the scientist as a reasonable man, always prepared to acknowledge and discuss an alternative viewpoint, applies only when the controversy at issue is considered as 'internal' to science as a whole. If the controversy is between a theory that is deemed to be scientific and one that is reckoned not to be, then the scientist takes exactly the opposite approach; that is, he simply disregards the 'non-scientific' alternative, and doesn't bother discussing it at all. For example, many religious 'fundamentalists' still refuse to accept the idea of evolution and hold instead to the view that all the earth's species of living organism were created, exactly as they are today, by some divine process. Yet, despite the existence of this anti-evolutionary camp, most modern texts on evolution do not even have an entry for 'creation' or 'creationism' in the index, let alone a chapter or two devoted to this view of the living world.

Needless to say, many creationists are not happy with this situation, and have clamoured to have their views heard, not only in all sorts of non-scientific arenas but also within biology itself. Recently, court battles have been fought in various American states, where the issue was whether the creationist view should be given equal prominence to the evolutionary view in biological education.

Now it seems to me that the creationist lobby had a choice of two quite different approaches. They could have argued that educationalists should give equal weight to opposing views of the world, regardless of whether or not either was deemed to be scientific. Alternatively, they could choose to accept the idea that scientists should consider only alternative *scientific* theories, but to argue that the creationist view did indeed fall into this category. As it happens, they chose the latter approach, which is why the phrase 'creation science' sprang into existence. I shall not use this label, because I think that it is false, for reasons that will become apparent later. But it is important to recognize that creationists have taken this view – that their ideas are indeed scientific ones – because whether or not that is true then becomes a central issue.

Is creationism science? This question begs another, namely: what is science? In other words, what distinguishing feature or features must some school of thought possess before we consider it to be a scientific view of the world and its adherents scientists?

We met this issue a long time ago – in Chapter 1 – where I noted that what distinguishes science from practical pursuits like car maintenance was its striving for generalization, whereas what distinguished it from metaphysical or religious beliefs was refutability. Clearly, the second of these contrasts is the important one here, so we need to examine this business of 'refutability' a little more closely.

The ideal biologist

Let us imagine an 'ideal scientist' and see how he behaves. In fact, we have to narrow things down a bit, since different ideal scientists behave differently! For a start we shall concentrate on the 'ideal biologist'. This means that we are examining a branch of science that has got past its early, pioneering stage, but has not yet reached the maturity of, for example, present-day physics. Even the 'ideal biologist' is a heterogeneous creature, for biologists in different fields operate in very different ways. The molecular biologist working out the DNA-sequence of a particular gene is engaging in an essentially 'atheoretic' pursuit, and we shall disregard him. Instead, we shall concentrate on a biologist who is concerned with the resolution of some controversy of general importance in a field where data of some kind can be brought to bear on the argument. Examples are population geneticists engaged in the selectionist/neutralist debate, and ecologists arguing over whether populations are usually regulated 'upwards' (via resources) or 'downwards' (via predators).

Our ideal biologist, so delimited, operates something like this. He is aware of two major alternative views of the world, both claiming to be general, which are incompatible. (Sometimes there are more than two, but that does not fundamentally alter our picture.) If two incompatible generalizations exist, covering the same range of phenomena, one of them must be wrong. Of course, *both* of them may be wrong, but we shall not consider this pessimistic, and rather uninformative, situation. Our biologist will probably feel himself drawn to one or other of the competing theories. The reason for this may be exposure to some earlier set of data that seemed to support 'theory A' rather than 'theory B'. On the other hand, it may be just an intuitive feeling that theory A is correct, or even that our biologist has a high regard for the originator of theory A.

Being an A-supporter, our biologist will probably design experiments intended to produce data that lend some support to theory A or, better still, that refute theory B entirely. You could say that this is a biased approach, and so it is, but this doesn't matter – partly because B-supporters are exercising their freedom to pursue the opposite bias. Both opposing camps will keep in touch with the accumulating data by reading the scientific literature. If the data seem ambiguous, both camps will probably continue to hold to their preferred theories. Even if the data seem, on balance, to favour theory B, without being conclusive in any way, our A-supporter may well continue in his belief in A's correctness. But if, later on, a really decisive new experimental approach produces results that refute theory A but are beautifully explained by theory B, our biologist will switch his affiliation and become a B-supporter. He may only do this after he has repeated the crucial experiment himself. Repeatability of results is very close to the heart of science, and our biologist would be foolish to be swayed by the results of a single experiment performed by a single researcher, however decisive its results might seem. But eventually, after one or several repeats, he is satisfied, deserts the camp he once defended, and regards the controversy as settled.

Although our biologist is indeed an 'ideal' one in many respects – really decisive experiments are rare in biology, for example – I have given him a number of decidedly realistic features. He operates partly on the basis of intuition, has an irrational attachment to the founder of his theory, is biased in many respects, and lets his sense of beauty influence his science. He is certainly not the caricature-scientist who spends his time objectively assessing a range of data painstakingly collected in an unbiased manner.

Now this may seem a dangerous course to adopt. If my intention is to contrast the biologist with the fundamentalist, it may seem perilous to imbue the biologist with too many 'human' characteristics, because this might be thought to blur the distinction I am trying to make. Yet the reverse is true, for one of the points that I want to make here is that *none* of the features of the 'caricature-scientist' are relevant. The only feature on which we need to concentrate is the willingness to change one's view when the evidence dictates it. Our ideal biologist possesses this feature, and that is all that is required to make him a scientist, and to consider his studies as part of the scientific endeavour.

This feature is conspicuously absent in any creationist I have ever met, and in any creationist publication I have ever read; and it is for this reason that we should not use the term creation science, nor allow creationism to be given any prominence at all, let alone equal status to evolutionary biology, in biological education. If people wish to propound a creationist message in another forum that is up to them; but masquerading as scientists is quite another matter.

Going back to predict the future

This point is of considerable importance, and I illustrate it with an example. For this purpose we must temporarily put ourselves back in the year 1960 or thereabouts. At that stage, we knew about the structure of DNA, the genetic code, and the basic essentials of protein synthesis; but we knew next to nothing about molecular evolution. We knew of course that any particular protein, such as haemoglobin, was found in a wide variety of organisms. We could thus speculate about what might be found if we were able to determine the precise amino-acid sequence of the haemoglobin molecule in lots of different species.

I do not want, here, to address details of *how* changes might occur between species if they have indeed evolved – we have done this already in Chapter 6. Rather, I want to inquire about the kind of variation in protein sequence that would be expected between species on the basis of (a) the idea that species evolved and (b) the idea that they were independently created. This is not an idle exercise, because much information on variation in protein structure is now available, and so we can contrast our 1960-based predictions with what we actually know.

The evolutionist's prediction is easy to arrive at. He takes the view that the same pattern of interrelationships that we have deduced from morphological studies should be apparent from the data on protein sequences. For example, haemoglobin from man and chimp should be very similar in sequence, while the comparison of human haemoglobin with that of a bird species should reveal many more differences. The two things – morphology and molecules – are seen as different manifestations of the same process, namely divergent evolution from a common ancestor. Thus the pattern of variation in proteins among species will have the same general form as the pattern of morphological variation, namely a hierarchical or tree-like pattern. Not only that, but the two trees should be reasonably congruent in terms of the species found on neighbouring branches. This is a highly dangerous prediction to make, because it is only one out of very many possible patterns and so has a very high likelihood of being refuted if the rationale underlying it is wrong. Such is the mental courage of the scientist.

The creationist's prediction is not at all easy to work out. If you put a hundred evolutionists in a hundred different rooms and asked them to write down what pattern they would expect, you would probably get a hundred predictions of a tree-like pattern. If you put a hundred creationists in a hundred different rooms you would certainly get several fundamentally different patterns – you might even get a hundred of them! Of the possibilities that occur to me, one is that the structure of haemoglobin would be identical in all the species we looked at. Another would be that the number of different amino acids in any interspecific comparison was always a prime number – or some other pattern that would reveal the design work of an intelligent Creator. The possibilities are endless.

Of course, we now know the answer: the pattern of interrelationship at the molecular level is broadly similar to the morphological. Thus the data corresponded to the evolutionist's prediction but not to any of the creationist ones that I have given.

Now this is not a fair comparison for several reasons, one of which is that I have devised both predictions despite my being an evolutionist, and the 'creationist patterns' I have suggested are not the only ones imaginable. Indeed, one could take the view that if living organisms have varying degrees of similarity at one level then they will have parallel variation in the degree of similarity at another level, regardless of whether or not they have diverged from a common ancestor. Thus we can have the same pattern of similarities as would have been produced by a tree-like process of divergence, without invoking the tree itself.

If you believe that one of the earlier 'creationist patterns' I gave was what our 1960-creationist would have devised, then his view is refuted. If he continues to adhere to it, he is not practising science, because he is refusing to accept evidence of a decisive kind. On the other hand, if you believe that it is impossible to say what the 1960-creationist would have predicted, and that a pattern inseparable from the evolutionist's must be included as one of his options, then the problem is not refusing to accept refutation but rather engineering things such that it can never occur. This can be done either by making your predictions broad enough to include any possible outcome or by ensuring that your predictions are always so close to those of your opponent that any time he's right, you're right too! *Sometimes* scientific theories make broad predictions, and *sometimes* opposing scientific theories make very similar predictions. But a body of theory which *always* engineers things so that it cannot be refuted is not scientific; and I think that creationism is precisely this. If any creationist thinks otherwise, perhaps he can make a specific and distinct prediction upon which future biological data can be brought to bear. If so, I suspect that his prediction will turn out not to be true, but it would have been a courageous attempt nevertheless.

Evolution and religion

It must be abundantly clear to anyone considering this matter that it is impossible to be both an evolutionary biologist and a religious fundamentalist. However, there is nothing in evolutionary biology that makes it incompatible with a more enlightened religious view – whether a Christian one or its equivalent in any of the world's other religions. One of the devious aspects of the creationist/fundamentalist lobby is that it attempts to see evolutionary biology as anti-religious. We can easily see why the creationist attempts to foster such a view. Acceptance of it would drive a wedge between biology and all religion, thus giving the creationist a host of apparent allies, rather than between creationists and 'reasonable people' (the latter including evolutionists and liberal religionists), which is where the wedge really belongs. In the end, evolutionary biology has little to say about religion of the non-fundamentalist kind – either for or against. It is possible to be an atheistic evolutionist, of course. But it is equally possible to be a religious one, and to view the origin of the universe itself as an act of divine creation. The origin of the universe is something which, unlike the origin of life, is outside the scope of science. Physicists can tell us a lot about conditions in the universe a fraction of a second after the 'big bang', but they can't tell us what happened before it, or what caused it. Whether one views the big bang as a divine act or a horrible accident is a matter of personal belief, not of science. Some of the day-to-day occurrences in human society seem to argue for the former view, others for the latter.

The dangers of creationism

As we have just seen, one of the main dangers resulting from the activities of the creationist lobby is the spread of the false view that evolutionary biology is anti-religious. This, however, is not the only danger. The integrity of biological education is threatened, one of the threats taking the following form. A biologist writes a school-level text which includes some material on evolution. The book is published and used in schools. A fundamentalist organization objects to the book's 'bias' and advises parents and schools not to buy it. Many of them heed this incompetent advice. Sales slump. In future, the publisher accepts only biology texts omitting evolution or giving 'creation science' equal coverage. This scenario – or something approaching it – has actually happened in some parts of the United States. The damage to biological education is considerable.

Evolutionists have not allowed these dangers to go unnoticed. An organization called the Association for the Protection of Evolution (appropriately APE for short) has been launched. Like any good evolutionary entity, it has begun to evolve – I recently noticed a letter to *Nature* which stemmed from a representative of 'Australian APE'! Whether this organization will ever have the resources and power at its disposal to defend evolutionary biology that the fundamentalist camp has to attack it is doubtful; but its very existence is at least reassuring – as well as a sign of the perceived dangers.

Finally, evolutionary biology is not the only branch of science that is threatened by the creationists. I recently saw in a copy of

The Plain Truth, published by the 'Worldwide Church of God', that this organization still adheres to the view that the human race was created 6000 years ago. This means that they reject well-accepted techniques in the physical sciences, such as radiometric dating, which allows us to estimate the age of fossil remains of man and other creatures. (Modern man, *Homo sapiens*, is at least half a million years old.) So, while the creationists falsely paint an anti-religious picture of evolution, the picture of creationists as anti-scientific is a true one. They may accept some areas of science that they don't feel threatened by – but their rejection of it goes far beyond evolutionary biology.

The search for the truth

When discussing the 'ideal biologist' earlier in this chapter, we ended with him deserting the theory he had previously held in favour of its rival. This is reminiscent of the baron who supports whichever king is in power at the time. However, the analogy is a superficial one, for the biologist's disloyalty to the theory he previously supported is due not to the urge for survival but to a loyalty to something deeper, namely a way of trying to find out the truth about the natural world. The biologist deserts his theory only when loyalty to it and loyalty to his methodology, which includes a critical assessment of the available evidence, become mutually exclusive. This supremacy of loyalty to a way of searching for the truth over loyalty to any particular theory is something biologists share with other scientists, and something that distinguishes the scientific approach from many others, such as that of the fundamentalist. However, the adherence to an entrenched viewpoint despite its incompatibility with the evidence is not restricted to those on the lunatic fringe of religion. Organizations in fields that have nothing whatever to do with religion also have vested interests which sometimes act against scientific progress in general, or against advance in some aspect of biology in particular.

One example of such an organization is Soviet communism, which, in the interest of maintaining its ideology, has propagated biological nonsense in various forms. One of its better-known idiocies in this area was to proclaim in its early days that the biological study of mental disorder was misguided because such disorders were caused by the capitalist system and would disappear under communist rule. Another scientific aberration associated with communist ideology, and one which is more directly relevant to this book, was Lysenkoism. Lysenko was a Russian geneticist working in the middle of this century and holding Lamarckian views which had long since been abandoned – by most of the biological community outside Russia, as well as by some biologists inside the country. Within Russia, Lysenko managed to dominate his field for a time not through being a better scientist but through his contacts in the party hierarchy, which enabled him to have scientific opponents sent off to Siberia when they were getting troublesome. The cooperation of communist officials in maintaining Lysenkoism in Russia can be attributed to the feeling that the Weismann/Mendel/Darwin 'axis', as described in the previous chapter, was too inherently capitalist, dominated as it is by the idea of survival of the fittest. Both Lysenko and Lysenkoism are now dead, so this particular attack on biological theory is over; but the continuing ability of a communist organization – or any other dictatorship – to interfere with biological progress which seems to contradict its ideology should not be underestimated.

I do not suppose that either religious fundamentalists or Soviet communists would be too enthralled at my lumping them together in this way. Other than a vested interest in certain ideas, and a consequent ability to disregard the evidence when it suits them, they certainly do not have very much in common. But it is precisely this one common characteristic that is relevant here and that is at the heart of all non-scientific attacks on biological theories. I hope our way of searching for the truth about nature will prevail against these attacks, but I would certainly not take that for granted. Various futures can be envisaged for theoretical biology: some of these are briefly considered in the final chapter.

15 WHAT THE FUTURE MAY HOLD

One of Bob Dylan's many sensible utterances was: 'He not busy being born is busy dying.' This applies not only to the physical and mental development of individuals, but also to scientific theories, both individually and collectively, and, for that matter, to the academic institutions from which they sometimes arise. (They don't all so arise: Mendel worked in a monastery, Darwin on a boat, and Einstein in a patents office!) Generally speaking, the twentieth century has been a good time for the development of theoretical biology, notwithstanding localized problems of Lysenkoism, fundamentalist subversion and Hitlerian book-burning. On balance, theoretical biology was very much being born.

Will this also be true of the twenty-first century and beyond? Of course, no one can tell. But rather than let the book end on that rather vacuous note, I would like to paint two pictures of the future – alternative theories of it if you like! I have deliberately chosen to paint extreme pictures, not because I think that either will come true, but because they identify the ends of a spectrum of possible futures, one of which will certainly occur.

A pessimist's future

At the worst end of the spectrum, we have the demise not only of theoretical biology, but also of humanity and possibly the biosphere as a whole, due to nuclear war. Nothing much can be said about this except that, like everyone else, I hope it doesn't happen.

Assuming the worst does not occur, what will become of our theories? One possibility is that the social systems of the countries in which most scientific research is carried out will cease to be conducive to our search for the truth. After all, the history of the human world is not written in the language of liberal democracies, which are a relatively recent event with a sometimes perilous-looking foothold in human history, comparable in a way to man's recent appearance in the history of life. It would be unwise to assume that we are on a stable path to progress and that the dark ages can never be repeated. It is hard to say whether the most serious threat comes from domination of the air-waves by rich fundamentalist organizations, 'people's revolutions' which lead to governments that regard pure science as unproductive (and people as expendable) or right-wing dictatorships hostile to anything that smells intellectual. All of these would be bad news for the theoretical biologist (and many of his other scientific colleagues) and, given any of these social backgrounds, sensible theories of life would hardly continue to be born.

Given a continuing biosphere, and a continuingly benevolent social system, what future is left for the pessimist? Well, nothing so far up his street as the above scenarios, but a depressing future can still be arrived at all too easily. Most western countries have recently seen a decrease in the number of openings for scientific researchers, and other academics, at universities and other research-oriented institutions. Those lucky enough to obtain a post have found it increasingly difficult to get funds for their research. Of course, you cannot stop someone having ideas by denying them funding, but the resolution of conflicts between different ideas requires experiments, and therefore money. By blocking funding, progress can be slowed to a trickle. And selective funding, for example of biotechnology over 'pure' biology, which is very much a current Thatcherian reality in the United Kingdom, can distort the pattern of progress that would have been manifested under a more egalitarian system of financial support.

If more enlightened future governments remove this financial stumbling block to the progress of theoretical biology, the pessimist can only resort to problems in the biological community itself. Like any other professional group, biologists are a heterogeneous bunch. The pattern of progress that we have seen – adherence to a particular theory until a better one comes along – depends on the interplay between two different biological mentalities. One is what I call the 'policeman' mentality, characteristic of those biologists who put their energies to the defence of established theories. The other is the 'rebel' mentality, possessed by an odd assortment of people, from crackpots to Mendels and Darwins, who are not satisfied with the prevailing view. Too many rebels would lead to chaos, too many policemen to stagnation. If the balance were to break down, then, theoretical biology would grind to a halt. Somehow, this seems not to happen; and this fact leads naturally into our alternative picture of the future.

An optimist's future

First of all, the optimist assumes that none of the above problems will happen. He then turns to consideration of the direction in which an unfettered theoretical biology might develop, and of the harmonious relations it might have with other fields of endeavour.

As regards its 'internal' progress, our optimist sees a few further theories being added to those we currently have, without causing incompatibilities to arise. He then sees a search for unifying ideas progressing to the point where our several theories are reduced to two – theories of regulatory and self-organizing processes, as described in Chapter 13 – and eventually one, the biological GUT, whatever it may turn out to be. Further, our optimist confidently predicts the discovery of life in distant solar systems that will fit into our theoretical picture regardless of any differences of chemical constitution it may exhibit from terrestrial organisms.

On the external relations of theoretical biology, our optimist sees a continuing compatibility with the theories of physical science; and a friendly mutual exclusivity of domain for biology and religion, in which investigation of the physical and spiritual aspects of life go their separate ways, with both acknowledging a lack of ability to refute each other's work and both being characterized by no desire to do so.

The actual future

What the next century holds for theoretical biology will probably be in between the two extremes portrayed above, and cannot, of course, be predicted in detail. But the degree to which it falls on the optimistic side depends, among other things, on you. The future, after all, is partly determined by events happening now, and everything that does happen now influences the future in some way, however small. Anyone who burns a book or refuses to think about a new idea helps take the future back into the dark ages. Anyone who observes the living world with curiosity and fascination, wonders how it works, and honestly assesses his own ideas about it, as well as those of others, helps to usher in a more optimistic future, and to contribute to the development of the theories of life to which this book has been devoted.

SUGGESTIONS FOR FURTHER READING

Since this book has roamed across the whole of biology, extracting theories here and there, and has even occasionally strayed out of biology into physical science, philosophy and religion, the texts to which the interested reader can turn to follow up the topics discussed are endless. They are also very varied in level, some being more suited to the general reader, others to the student of biology and the professional biologist. Those titles given below are just a small selection out of the vast range of relevant texts, chosen because their authors were particularly concerned with the development or explanation of biological theory, with major biological discoveries, or with the nature of scientific progress in general.

For the 'interested outsider' I would recommend the following:

The Ghost in the Machine, by Arthur Koestler (Pan Books, 1970)
The Theory of Evolution, by John Maynard Smith (Penguin Books, 1975)
The Art of the Soluble, by Sir Peter Medawar (Methuen, 1967)
The Double Helix, by James Watson (Penguin Books, 1970)
Ever Since Darwin (and later books in the same series), by Stephen Gould (Penguin Books, 1980)

For the actual or aspiring biological theoretician I would recommend two 'batches' of books. First, a heterogeneous group of texts (group A below) each of which is particularly outstanding in some way. Second, a group of books (B below) dealing with Darwinian and developmental approaches to evolution, and their compatibility. This second group is included because, as mentioned in Chapter 13, the degree to which these approaches are indeed compatible is one of the major remaining puzzles in theoretical biology.

GROUP A

The Structure of Scientific Revolutions, by Thomas Kuhn (University of Chicago Press, 1970)
On Growth and Form, by D'Arcy Thompson (Cambridge University Press, 1917)

Towards a Theoretical Biology, edited by C. H. Waddington (Edinburgh University Press; four volumes 1968–72)

GROUP B

Mechanisms of Morphological Evolution: A Combined Genetic, Developmental and Ecological Approach, by Wallace Arthur (Wiley, 1984)

Ontogeny and Phylogeny, by Stephen Gould (Harvard University Press, 1977)

Epigenetics: A Treatise on Theoretical Biology, by Søren Løvtrup (Wiley, 1974)

Evolutionary Theory: Paths into the Future, edited by J. W. Pollard (Wiley, 1984)

Embryos, Genes and Evolution: The Developmental Genetic Basis of Evolutionary Change, by R. A. Raff and T. C. Kaufman (Macmillan, 1983)

The Strategy of the Genes, by C. H. Waddington (Allen & Unwin, 1957)

Finally, for *everyone* who has not read it, I could not recommend too strongly Charles Darwin's *Origin of Species*, first published in 1859 (and available in paperback: Penguin Books, 1968). It is often referred to as the most approachable of the major works of science, and rightly so in my opinion. It is readable and informative to layman and professional alike, which makes it an enviable piece of writing as well as a biological classic.

INDEX

FOR THE BEST IN PAPERBACKS, LOOK FOR THE

A CHOICE OF PENGUINS AND PELICANS

Dinosaur and Co Tom Lloyd

A lively and optimistic survey of a new breed of businessmen who are breaking away from huge companies to form dynamic enterprises in microelectronics, biotechnology and other developing areas.

The Money Machine: How the City Works Philip Coggan

How are the big deals made? Which are the institutions that *really* matter? What causes the pound to rise or interest rates to fall? This book provides clear and concise answers to these and many other money-related questions.

Parkinson's Law C. Northcote Parkinson

'Work expands so as to fill the time available for its completion': that law underlies this 'extraordinarily funny and witty book' (Stephen Potter in the *Sunday Times*) which also makes some painfully serious points for those in business or the Civil Service.

Debt and Danger Harold Lever and Christopher Huhne

The international debt crisis was brought about by Western bankers in search of quick profit and is now one of our most pressing problems. This book looks at the background and shows what we must do to avoid disaster.

Lloyd's Bank Tax Guide 1986/7

Cut through the complexities! Work the system in *your* favour! Don't pay a penny more than you have to! Written for anyone who has to deal with personal tax, this up-to-date and concise new handbook includes all the important changes in this year's budget.

The Spirit of Enterprise George Gilder

A lucidly written and excitingly argued defence of capitalism and the role of the entrepreneur within it.